LIFE ON OTHER PLANETS

LIFE ON OTHER PLANETS

A Memoir of Finding My Place in the Universe

AOMAWA SHIELDS, PhD

VIKING

VIKING
An imprint of Penguin Random House LLC
penguinrandomhouse.com

Copyright © 2023 by Aomawa Shields
Penguin Random House supports copyright. Copyright fuels creativity, encourages diverse voices, promotes free speech, and creates a vibrant culture. Thank you for buying an authorized edition of this book and for complying with copyright laws by not reproducing, scanning, or distributing any part of it in any form without permission. You are supporting writers and allowing Penguin Random House to continue to publish books for every reader.

A portion of this work previously appeared in "Universe: the Sequel" from *She's Such a Geek: Women Write About Science, Technology, and Other Nerdy Stuff* edited by Annalee Newitz and Charlie Jane Anders. Originally published by Seal Press in 2006.

Grateful acknowledgment is made for permission to reprint the following:

"Watching the moon" from *The Ink Dark Moon: Love Poems by Ono no Komachi and Izumi Shikibu, Women of the Ancient Court of Japan* translated by Jane Hirshfield with Mariko Aratani, translation copyright © 1986, 1987, 1988, 1989, 1990 by Jane Hirshfield. Used by permission of Vintage Books, an imprint of the Knopf Doubleday Publishing Group, a division of Penguin Random House LLC. All rights reserved.

Excerpts from "Space Was Her Second Act: As Spitzer Takes Its Final Bow, Astronomer and Actress Aomawa Shields Thanks Spitzer For Her Big Break" by Aomawa L. Shields used with permission of The California Institute of Technology.

Lines from "Smelling the Wind" from *The Collected Poems of Audre Lorde* by Audre Lorde, published by W. W. Norton & Company. Used with permission of W. W. Norton & Company.

Library of Congress record available at https://lccn.loc.gov/2022043791

ISBN 9780593299180 (hardcover)
ISBN 9780593299197 (ebook)

Printed in the United States of America
1st Printing

DESIGNED BY MEIGHAN CAVANAUGH

While the author has made every effort to provide accurate Internet addresses at the time of publication, neither the publisher nor the author assumes any responsibility for errors or for changes that occur after publication. Further, the publisher does not have any control over and does not assume any responsibility for author or third-party websites or their content.

For my little girls

Aomawa as a child

looking up at the stars

and

Garland-Rose

A universe of possibility

No star is ever lost we once have seen,
We always may be what we might have been.

ADELAIDE ANNE PROCTER, 1859,
"A LEGEND OF PROVENCE"

Watching the moon
at dawn,
solitary, mid-sky,
I knew myself completely,
no part left out.

IZUMI SHIKIBU, TENTH-CENTURY
FEMALE JAPANESE POET
JANE HIRSHFIELD AND
MARIKO ARATANI, TRANSLATORS

CONTENTS

Prologue *1*

BEGINNING

The Big Bang *7*
Expansion *18*
Jazz *30*
High Flight *36*

CHOOSING

Physics *47*
Research *57*
Tide *68*
Migration *79*

Acting 86

Home 92

Temping 96

Break 106

Recovery 110

Pluto 116

Stars and Life 119

Hair 124

RETURNING

Solar Cycle 131

Snowballs and Life 145

The Qual 153

Ice 164

The Weather on Other Worlds 172

Gravitational Interactions 180

Congratulations, Dr. Shields 185

MERGING

Different Stars 193

Harvard 199

TED 208

Kids 211

Exploding Stars 216

A Bird's Wing in Taos 220

Evolution 224
Pregnant 234
Welcome to the World 238
Plenary 249
NEOWISE 256

RISING

I Never Thought I Was Only Located Here 265
Aliens 267
Congratulations, Professor Shields 274
First Day of Class 283
Twelve Things Going through My Mind during an Average Minute of My Life on Retreat in Ojai 290
Destiny 292
Home 2.0 303
Journey 307

Acknowledgments 317
Mental Health Resources 323
Further Reading 327
Bibliography 329

Arecibo Observatory Staff, Puerto Rico, 1996

LIFE ON OTHER PLANETS

Prologue

Everything has a beginning, a point when a single spark ignited everything that was to come. Then life separates into two regimes: before the spark, and after.

It's hard to think back to how it all began. Looking around—at my skin, the weather, the piles of papers on my desk, the shelves of books about astronomy and planetary climate—it seems like it was always there. But it wasn't.

In one of the textbooks that I use to teach the course Life in the Universe, an analogy is given: there are as many stars in the observable universe as there are dry grains of sand on all the beaches in the world. Even as an astronomer and astrobiologist, this shocked me. It drove home the point of just how big *big* truly is. This is something we're always trying to get students to grasp in introductory astronomy courses—the sense of scale when we talk about astronomical distances and phenomena. We say things like, "If we put the Sun in Washington, DC, and shrank distances between it and everything else down by a factor of ten billion, where would the nearest star to the Sun be?" The answer: San Francisco, California.

I always get a big reaction in class when I reveal the answer. Even after wrapping their minds around shrinking the distance down by a factor of ten billion, a number itself so hard to comprehend, the nearest star to the Sun would still be across the entire country? It stuns my students, and widens their eyes. I pause for a moment at the front of the lecture hall, staring into their faces, and let it sink in for effect.

On January 20, 2009, Barack Obama became the first African American president of the United States. In the fall of that same year, I started a PhD program in astronomy and astrobiology. This was my second attempt at a doctorate. I remember feeling nervous, scared, and also proud. Together, President Obama and I could do it. *Yes, we can.* We could inhabit spaces dominated by white people—white *men*—and excel. We could blaze the trail. We were pioneers. President Obama was my cheerleader from afar, even if he didn't know who I was. Somewhere across the country, in the highest rank in the land, or in the air traveling to distant countries to broker peace and soothe nations, was someone who knew how I felt, probably felt the same way some of the time, and was doing it anyway. His becoming president was a lighthouse in a dark sea to me, signaling with regular frequency the existence of a shoreline and sure footing close by. The inaccessible world of astronomy felt no longer closed to me if the most distinguished appointment in the country was now occupied by a Black person.

Returning to astronomy with my whole heart meant leaving something behind—a career in acting. Acting and astronomy lit up my life in different ways, and each seemed to demand all of me. When I left astronomy to pour myself into acting, I thought it was for good. But dreams don't die. If left unpursued, they recede, and then lie dormant. But they persist, nudge, sometimes elbow you right in the gut and demand to be acknowledged. That's how it was for me. I fell in love with something, and I never really fell out. But it wasn't the only thing. This made life complicated. Is this your story too?

There is a type of planet that is very different from the Earth. This

PROLOGUE

type of planet rotates so slowly that the length of its day is the same as the length of its entire year! On one side of the planet—the side facing the star—it's always daytime, and its sun stays fixed and relentless in the sky above at high noon. On the other side, it's perpetually night. Can you imagine the Sun never rising where you live? Never setting? *Ever?* This happens at the poles of the Earth at certain times during the winter and summer months: the Sun never dipping above the horizon, or never leaving. But eventually, the Sun returns to view. Eventually, the Sun recedes.

Not on this planet. On this planet, the day side, baking under the eternal noon of a different sun, could be too hot to support life. And the frigid night side, bereft of the warmth from its star, could be too cold. But there is a place that life might have a hope of existing on this planet: along the dividing line between the two halves, a line called the "Terminator." (Yes, this makes me think of the action-packed films too. And indeed, a movie or TV show is often the catalyst for much in my life. You'll see what I mean. But this is a different kind of terminator. One far more neutral.)

My research team calls planets with climates amenable for life only along this boundary line between two inhospitable extremes "Terminator Habitable." We weren't sure if they could exist; we did a lot of work to figure out that they can. I'll tell you about the kind of work soon. First, though, I had to ask the question, "Could planets exist with this type of climate?" I started there. A sense of wonder is always a good place to start.

We start out in this life as babies looking out at the world in wonder. We discover things we like, things we don't like, and things we love. We make a choice, and try to create a future with it that fills us up inside. Along the way, life happens. We might abandon some of those loves—to support ourselves and our families financially, or because what we wanted to do doesn't seem possible, because it's too hard or it's too late or it doesn't make sense, no one's ever done it before. Does any of that ring true for you? I wrote this book for you, so that you know that it isn't too late to have what you really want in this life, whatever that is, and whatever you perceive as obstacles blocking your path. You may feel like a rare, magical unicorn at what you are trying to do with the background and set of circumstances that you currently have. But you are not alone. There are other magical unicorns in the meadow.

I started my life with one thing: science. Astronomy, to be specific. And I dove into it. Then I found something else I liked: the arts. Acting, to be specific. So I dove into that instead. Neither one by itself felt fully right. I spent a lot of time contemplating how to make them both work. I spent more time wondering if it was even possible to. I'd never seen it done. But I had a hunch that there was no one way to do it, and possibly even more ways than I could imagine. So I took a risk that if the interface between the two worlds felt like home, there would be air there to breathe. I was right. I see it confirmed in these "Terminator Habitable" planets. The "in-between" can be sustainable. There are all kinds of planets out there. The universe is a very big place. There is room for it all.

June 2022

BEGINNING

The Big Bang

The sky was my first love. As a child, I was always looking up. So much so that I often bumped into things, running into poles and buildings on the street. There was so much going on up there, and it seemed far more interesting to me than walking around with my eyes pointing straight ahead, unaware of it all.

As a kid, I wanted to be a lot of things when I grew up. Did you too? For a while I could only think about becoming a Dallas Cowboys cheerleader. I don't remember when that dream formed itself in my mind. It wasn't like I watched a lot of football as a child. But back in the '80s, the Dallas Cowboys Cheerleaders were larger than life. There was probably a Super Bowl that I watched and saw them jumping and dancing and kicking up their heels high as they cheered in those white ten-gallon hats and blue-and-white vests and short skirts with tassels everywhere. And of course, white cowboy boots. I wonder how comfortable those were to dance in. I wanted to be like them—beautiful and white and always wearing those outfits.

When I was seven or eight years old, I would put the neck of a sweater over my head so that the rest of the sweater fell down my back like a cape and only my face was visible through the opening. I'd go to the playground and get on the swing with my sweater on my head. I would build momentum, going higher and higher in the air until it felt almost like I was flying. The sweater on my head felt like hair—flowing, long, fine, golden, "white girl" hair. I'd fling that sweater around my head like I was in a Pantene commercial. I wanted hair that flowed down my back, rather than hair that stood up dark and kinky and coarse, that stopped my fingers as I tried to comb through it. The Dallas Cowboys Cheerleaders were the epitome of "all-American" beauty. And that's what I wanted to be: a beautiful all-American girl.

My mother was appalled. A PhD in music theory and composition, a professor at a progressive college in Amherst, Massachusetts—this was not what she had raised her daughter to aspire to be. She had high hopes for me. All she could do was pray that my aspirations would change with the seasons, if not sooner.

They did. I decided one day, probably after visiting the school administrative office on an errand for my teacher, that I wanted to be a secretary. I was fascinated by the number of office supplies that the desks of secretaries contained—ledger books and notepads, files that got inserted into cabinets and other shelving units. There were paper clips and binder clips, and thumb tacks stuck pinching papers against cork bulletin boards. I loved the idea of taking notes. What were all of these communications about? I wanted to know.

My mother winced again. Now of course, neither cheerleader nor secretary is by any means a shameful occupation. But you have to understand where I came from. In my home, education was prized almost above everything else. Education was how you got better jobs, houses, bank accounts with nonzero balances. Not many people in my family had graduated from college. My mother going all the way to getting a doctorate in something had brought up lots of feelings in extended family members. At least, this is the story I heard growing up. How my mother was treated as if she considered herself too good for them. Without much in the way of open communication, the relationships withered and fossilized. It was me, my mom, and my grandmother for most of my childhood, with male figures who came and went—father, then stepfather, then later, when I went to college, another stepfather.

My mother and father were professional musicians. They created my first and middle names out of vowel sounds. Ah-*oh*-muh-wah. Lah-*lah*-hee-ay. No glottal stop at the back of the throat between the sounds (you might need to look that up—it's where you stop the flow of air through your throat by closing your glottis, which lies at the back of it). Just nice and fluid. My name was a chant. *Aomawa. Aomawa Lalahie Aomawa. Aomawa.* I wish I could sing you the notes, but maybe it's better for you to have your own version of it in your head. My parents created a meaning for my name too. *Aomawa Lalahie* together means "Spiritual Strength." Not bad.

Throughout my life I have been asked if I have a nickname, usually after being asked, "Is that African? It sounds African." Sometimes people guess other origins, like Hawaiian, or Native American, or Japanese. I like this, because it seems less like they are making assumptions based on my skin color. The best question I've gotten is "Where does your name come from?" Pure question, no assumption.

"My parents made it up. They're musicians. They created it out of vowel sounds."

I watched their eyebrows lift toward the sky.

The only person who calls me anything other than Aomawa is my mother. She calls me Aom, but pronounces it like *Om*. I've always loved her nickname for me. Later, when I started doing yoga, I loved the *Om* in my name even more.

I didn't love my name at first, as a kid. I wanted to have a "normal" name, like "Sarah" or "Jennifer" or "Rebecca," a name that was easy to say, that didn't make the teacher pause before saying it at the start of each new school year; at the start of any new anything, really. What followed was always a butchering of it—uh-*mow*-uh or *Ay*-oh-*mow*-uh or Ad-manda (attempting to find something familiar in my name). Other kids weren't kind.

It took years, but somewhere around junior high school, it started to be cool to stand out, to have a name that no one else in the world had—I embraced it and never looked back. I still always know when someone has gotten to my name while reading a long list of names for the first time. I listen for the pause, and I'll often beat them to it, saving them from that first pass by telling them how to say it right from the start. Much later, I'd even start out my science talks with a pronunciation key and lead the audience in a group practice of how to say my name. I took ownership of it.

My parents' marriage lasted only two years. Then they went their separate ways, my dad plunging himself into his music and travel, my mother going on to graduate school, a faculty job, marrying again, having my brother, ten years my junior, and later, after a new marriage, my sister, twenty-one years younger than me. Though decades have passed, their band reunited recently. My mother and father toured Europe in 2010, and then again in 2019. They recorded an album in London. They were working on a new album as I wrote this. My father has assembled a new version of the band and is still touring. The thread of creation through the arts that brought my parents together, and brought me into existence, still runs thick and strong through our family.

I grew inside of my mother as she played and danced onstage with my father. In pictures she wears gorgeous multicolored outfits with belly out and bulging, flute in hands against mouth. All that music is in my blood.

I am thinking about a time when I was ten years old and we lived in La Jolla, California, while my mom was pursuing her PhD at UC San Diego. We lived in graduate student housing, in a two-bedroom ground floor apartment. My grandmother and I shared a bedroom, and Mom had the other one, which also held her music composition table, and later, a crib when my brother was born. But that would come later, after my stepfather John, who is dead now, entered our lives. At this moment in my memory, it is just me, my mom, and Grandmother. Grandmother is sitting in her rocking chair in the corner of a softly lit living room. She is praying. She isn't praying for a relative to emerge safely from surgery, or for an end to some war on the other side of the world. She isn't even praying for food for us for the next day. She is praying for my mother to come back safely from the grocery store.

We were a house full of women who were self-sufficient and independent. I never knew my mother's father, but he cast a big shadow that filled every room. A truck driver, a drinker. He died from a heart attack at the age of forty-two, when my mother was in boarding school in Putney, Vermont. I would hear stories of when he and my grandmother would "fight," meaning he would get drunk and beat the shit out of her. I learned this years later. How my grandmother told my mother to always be nice, so that relatives would take her in if something happened to her. The undercurrent of fear crept into my mind and behavior too. I always called if I was going to be late coming home from a friend's house. I never fell asleep when my mother and I were driving home at night. Grandmother told me not to. Something bad might happen. Mom might black out at the wheel like Grandmother had long ago. Our fear was electric. It powered us. It also got things done. It kept the lights on after the third notice of delinquency.

My mother shopped at high-end cosmetic counters. I loved the makeup and would run my delicate fingers over her cosmetic bags and shoeboxes full of blushes and eyeshadow palettes and brushes. I learned the brands that were the best—Shiseido, Shu Uemura, Lancôme, Estée Lauder. In my teens, the Clinique 3-Step skin care system was my introduction to this kind of makeup and facial care. I was instructed to cleanse with the bar first, then rinse. Pat dry, then use a cotton ball to sweep the toner across my face. It stung a little bit, but also felt refreshing. I learned this was what an astringent was. It removed impurities. Impurities were dirt, grit, sweat—everything that attached itself to my face as I flew through the day. Lastly, the yellow moisturizing lotion. My mother taught me to be patient with each step in the process. It was how my skin would stay clear. I did as I was told. I pored over her *Elle* and *Vogue* magazines. She wasn't above the *Redbook*s or the *Marie Claire*s, especially if there was a woman on the cover she admired, like Isabella Rossellini or Linda Evangelista or Christiane Amanpour. Their high cheekbones and long lines, faces from another time, heritages steeped in the combined traditions of multiple cultures—women not easily figured out—were my mother's thing. I found young faces like mine in the pages of *Seventeen* and *Glamour*. The two of us ogled the glossy pages of these magazines, gently turning each page as if it were an ancient Hebrew text unearthed from a dusty tomb in Acra. These were our jewels, and our maps to beauty and glamour.

I wondered why money never seemed to stay long in our home. My mother was a professor, after all. At the beginning of the month it rolled in. Dinners out followed, along with new clothes and makeup for us, toys for me and my brother. By the end of the month, we were stretching our pantry into a taut rubber band. My mother's colleagues planned vacations to islands and other countries. On my mother's sabbaticals, we stayed home. I never made the connection until my early thirties, when I found myself doing my own kind of math at high-end department store cosmetic counters, in CD stores, at Beverly Hills spas that served celebrities. As the Earth continued to spin, never stopping to take a breath, something was missing, and I was afraid to sit still to figure out what it was. While the sky full of stars rotated above my head through the seasons, I was afraid to look in the mirror. I might notice that I didn't know the person staring back at me.

I had been told since as far back as I can remember how much I had been wanted. My mother couldn't wait to have me. I had been wanted in the world, from the start. I knew that not everyone had been told that. To start out in this world as someone who was loved meant something.

But I didn't know who that person was that was loved. I knew who I needed to be, for my mom, for my grandmother, so that everything would be okay for them, and for me too. But I'd lost myself. And I didn't want to see that then. So I shopped instead. I thought that by bringing more things into my life I would compensate for what was missing. I hadn't thought about something. When the gas tank is empty, filling up the car with everything else you think you need besides gas won't solve the bigger problem. You're still stuck.

My mom might have been one of the few in her family to graduate from college, but the distinction of the first in the family to *go* to college belonged to her mother, my grandmother Delphine, who lived with us after my parents divorced when I was two and we'd moved south from Oakland.

My grandmother loved numbers. She majored in mathematics in the 1930s at Tennessee State University, a historically Black college. She was the only woman in her class. As sparse as Black women are today in math and science programs, in the 1930s they were virtually nonexistent. She raised chickens to put herself through school. She kept a coop in her backyard and sold the eggs. One day the chickens escaped from the coop. That was the end of her college career.

My grandmother was as strong as they come. She went on to weather a lot—a stillbirth, the violent drinking and truck-driving husband, and keeping afloat a hair salon in Buffalo, New York. She didn't back down from many things. This is why I know things had to have gotten bad for her to leave that math program. As a Black woman sitting at the front of a lecture hall full of Black men, she was never called on. Instead, she was ignored, like a shadow. When she lost her coop of chickens and the financial support they brought, it was the final blow. But she was always trying to get back. When I knew her—that is to say, when our life spans overlapped—she had stacks of textbooks on her bedside table. Math textbooks, accounting textbooks. She went to night school. She would often sit in her rocking chair with a pencil and scribble numbers onto the back of an envelope. She never missed the televised California lottery announcements. Even when they didn't match the numbers she'd chosen, which was every time, she recorded the sequence of numbers in her notepad. She combed through those sequences later, looking for patterns. She watched *The Price Is Right* on weekday mornings. Not for fun: she was doing research. She would guess the prices of the living room sets, Jacuzzi tubs, and four-wheel-drive vehicles that would emerge from behind the moving walls on the show. Amid the cheers and oohs and aahs of the studio audience, Grandmother was doing the math in her head. She might be on the show someday, and she wanted to be ready. Learning how to guess a number that was close enough to the actual price of a consumer good without going over it was a skill—one that didn't require the bachelor's degree that she didn't have. Instead, it used her real-world experience shopping for her family over the years.

Grandmother worked with numbers at Naval Air Station Miramar in San Diego. She was a civil servant, not a member of the military. She was single-handedly responsible for my winning the school cookbook sale. She took the order form I'd gotten from school and brought it to her department at Miramar. I sold over one hundred cookbooks and won the grand prize balloon sticker with my name on it, affixed to the top of the class cookbook sales tree. And, most important to me, a gift certificate for books.

Miramar was the home of the Blue Angels, the renowned air force aerial flight team. We went to Miramar on weekends to watch them fly. We would load up coolers and blankets and lawn chairs and line up for miles. Then we'd crane our heads back in the hot sun to peer up into the sky and watch these cobalt and mustard-yellow planes cut through the burning blue in death-defying, intricate formations, coming within yards of each other to make bow ties and diamonds in the sky.

I would watch the sky's piercingly clear crystal blue, punctuated with fluffy white streaks of contrails. That blue was a feeling. I would lie back on the soft checkered blanket and feel as I watched. I felt the tiny hairs of my arms warm and sink closer to my skin and release, relaxed by the temperature of the Sun. I didn't know then how truly unique the Sun was, its yellowish-whitish glow so uncharacteristic of the vast majority of stars in the night sky, which are mostly redder and cooler and smaller. Then, I simply felt comforted by the blue as I watched the planes dive and curve and make patterns and shapes that I could hold out my fingers to trace. The blue was for me, so that I could see these planes in the sky, so that my family could live and work and enjoy something special on the weekends with the rest of San Diego. No place else existed. In no other town or state or country was the blue *this* blue.

I wouldn't truly understand how the sky got to be that blue until graduate school. How it wasn't from the sky reflecting the color of the ocean. The blue is everywhere—over the Sahara just as well as over the Mediterranean. How the sky is an atmosphere. How the tiny pieces of that atmosphere are really good at scattering blue light from the Sun much more than any other kind of light. How humans have only known this for a little more than 150 years. All that time before, not knowing what caused the color over our heads. When I found out, the new knowledge was an emerald jewel I kept inside my body, warm and glowing near my heart, with a thin silk thread connecting it to my mind. I could never not know this now.

Miramar was also the home of the original TOPGUN fighter pilot training school that inspired the 1986 movie. I still remember the moment in that movie when Kelly McGillis's character, Charlotte Blackwood (call sign "Charlie"), was introduced for the first time for who she really was, the morning after the night at the bar when Tom Cruise hit on her and thought she was just some random chick. The next morning in flight class, wearing sheer black pantyhose and pumps, she walked down the aisle between the two groups of male fighter pilots in their jumpsuits. Michael Ironside said in his gravelly voice, "She has a PhD in astrophysics, and she is also a civilian contractor, so you do not salute her." Charlie got to the front of the class and whipped her head around, her Aviators with the mirrored lenses shining, her wavy blonde hair catching the sunlight. I remember thinking, *Hell yeah, astrophysics. Sign me up. I want to be a badass like her.*

A year later, when I was twelve, my school class was shown the movie *Space Camp*, about a bunch of kids who accidentally get launched into space while they're at space camp over the summer. As I watched, I got really still. You know, the way you get when everything about something is you—all for you—and you know it, even if no one else does yet. I had always wanted to go to space camp, but we could never afford it. Still, the kids in this movie—a young Joaquin Phoenix, Kelly Preston, Lea Thompson, Tate Donovan, and Larry B. Scott (the Black guy from *Revenge of the Nerds*; I had to look him up. Why didn't he get to have his own romantic love interest in the film?)—were me. If they could get launched into space, then I could too.

Everything as we know it and see it today came from a tiny, pregnant point of nothing. In nearly the smallest fraction of a second, 13.8 billion years ago, this point exploded into a sea of darkness, expanding outward fast and wide in all directions. First hot and dense, as energy spread out, stretching, things started to cool, building an outline of structure for matter and form to develop. That is how our universe began. That is how the promise of everything that came after, including you and me, was born. Once the universe became cool enough for gas to condense, the gas flattened and began to spin. The first stars formed in these clouds. Small debris in these gaseous disks found other debris in the darkness, pulled together by mutual affinity. Once large enough, the weight of their own gravity began to tug and pull and spin them into spheres, where they became planets and moons. This is how things grow in the universe—by gravitational attraction, and the accumulation of larger and larger bodies to form something new. But first these bodies have to find their kindred in the distance between to reach their full potential. Accretion is a mysterious process. Bodies wander aimlessly in the darkness, until something happens. In a flash of brilliant light from a projector screen burning through my eyeballs, I was home.

Expansion

My mother purchased a set of *World Book Encyclopedia*s. I fell in love with them, how I could pick a volume for any letter and look up subjects in it that started with that letter. I could read and learn all about those subjects. And see pictures too. *World Book Encyclopedia*s weren't like *Encyclopedia Britannica*s, full of tiny text on rough paper with a few sketches in gray scale. They were printed on shiny pages that felt smooth underneath my small fingers, and in color. When for a brief moment in time I wanted to be an orthopedist, I got out the *O* encyclopedia and looked up *orthopedics*. I read all about bones. Bones went throughout the entire human body. There were so many of them—206!—beneath the surface of the skin, with tissues and cartilage threading between.

The afternoon that I saw *Space Camp* at school I ran home and got out the *A* volume of my *World Book*s and looked up *astronomy*. The pictures stopped me. Stars and colors—pinks and blues—in space. Wispy threads of what looked like fabric wafting through the blackness. I would learn much later that these shapes were something called nebulae—clouds of ionized (aka "excited") gas, the electrons getting more energized as a result of starlight shining on them. And since energy is conserved—what goes up must come down—those excited electrons calm down like children after a sugar high. The electrons release that energy as tiny particles of light called photons, with color and brilliance. I would learn later how those nebulae are nurseries. Stars are born in these dense clouds of excited gas. But here, now, looking at the pictures for the first time, they were simply pretty and powerful and I wanted to see more.

I looked up *astronaut*. I saw a picture of a man in a large white puffy suit, a helmet on his head with a clear visor in front. He was floating in space, attached to a chair called a Manned Maneuvering Unit, or MMU. I saw a picture of a large structure in space—one of the early Russian space stations—where people lived and worked. And not far below it was the bluish glow of something large and curved. I read the words *low-Earth orbit*. That larger thing below was the Earth! Where I was standing right at that moment. Years later, I would see pictures of the Hubble Space Telescope taken by space shuttles separating from it after servicing visits, the backdrop of the curvature of the Earth, the blue sky shading into black. Those people—astronauts—were not on this planet. They had left it, escaping the pull of gravity. Nothing that happened on the Earth during the time they orbited this planet would affect them. An asteroid could hit, a bomb could explode, and there they'd be, suspended, untouched, unfazed by everything we had grown up to believe mattered.

I wanted to be up there.

I plotted my entire career trajectory that afternoon. I would get a PhD in astronomy from the best science school in the world—MIT. Then, after a few years of professional experience, I would apply to NASA's astronaut candidate program.

I devoured television shows and movies about space and science fiction. My family watched *Star Trek: The Next Generation* every week. *The Abyss*. Later, *Apollo 13*. The *Battlestar Galactica* reboot. I salivated over the stories, putting myself in the characters' places, living their precarious lives in the deep unknown reaches of existence.

When I was fourteen, I was at the top of my math class, along with a boy I had a crush on. I didn't want to miss out on anything he was doing in school. I found out he was applying to Phillips Exeter Academy, an elite prep school in rural New Hampshire. So I applied too. When I went to visit, I found out that they had their own observatory. That's what got me.

I went to Exeter in the tenth grade and forgot about that boy. We both dissolved into our own lives, dictated by the location of our dorms. He had gotten into a dorm at the far outskirts of campus. I'd been assigned to Dunbar—*the* dorm to be in as a girl. It was central and full of blonde rich girls who dressed like they'd just gotten out of bed. They knew how to have fun. My life opened up, into dances and lip-sync contests and pining over other boys.

Every day I walked into the dining hall and had my daily inner conflict: Do I sit with my white dormmates, or at the Afro-Latino Exonian Society table, where most of the other Black and Brown students sat? Some days I steeled myself and, tray shaking slightly, walked over to the table of Black and Brown faces. I asked timidly, "Can I join you?" Someone nodded or said a quick "sure." I sat there in silence mostly, listening to whoever was telling a story or a joke. I smiled and laughed at the right times. I didn't feel comfortable. But I felt relieved. I wasn't denying my heritage or my people today. I was sitting with them. No one could call me a "Klondike" or an "Oreo." No one could tell me that I didn't know that I was Black today.

On other days, I wanted it to be easy. I'd walk in with my dormmates, get my food, follow them to a table on the other side of the dining hall, and continue the conversation. I'd be the only Black face there. When my roommate and best friend, Yvonne, a Mexican American who didn't seem to struggle with the same conflict (she sat at the Dunbar table or with her boyfriend), sat with me, we were a Black and a Brown face in a sea of white and blond and sandy-brown hair.

I had more to say at these meals. I lived with these girls. We had boys to talk about who visited the dorm. I usually had a crush on one of them. And he was usually white. Or better yet, someone who was white but acted like many of the Black students who hung together. He listened to the same music, used the same slang. This kind of boy was attractive because he could weave in and out of both hostile territories without detection. I thought that since he was familiar with both of my worlds, he would understand me. Again, I was doing the wrong kind of math. I was too young to know how many other variables needed to be part of that equation.

We had boys in common, but the lives of most of my dormmates were still different from mine. Few, if any, of them had ever feared the electricity being turned off in their homes. Few, if any, were on financial aid, which meant they didn't have to have a job on campus, like mine giving tours to prospective students and their families for the admissions office. Few, if any, had gotten a letter at their home telling their parents that if the tuition wasn't received by the end of next week, the administration would be happy to help them find another school to attend. Few, if any, had escaped out the back porch door when an argument went bad, mother and grandmother in tow, leaving a defeated stepfather in the living room to break toys and wonder how things had ever gotten to this point. Few, if any, worried about dressing nicely for class to make sure the school knew they were grateful just to be there, to be allowed to study in such a place, next to children of parents who occupied the top 1 percent of earnings in any country. Few, if any, worried about what table they should sit at in the dining hall every day, and who might judge them for that choice. Few, if any, worried about such things at all.

As soon as I could, I took astronomy courses at Exeter. My instructor, Mr. Harper, was a cheery older man from New Zealand with a full head of short, sloppy, salt-and-pepper hair and a small pair of spectacles that he wore at the end of his nose. With a broad face, a wide, boyish smile, and lightness in his feet, he seemed to be *so* excited to be there in the physics classroom, teaching us about the universe. He told us to start keeping astronomy logbooks—notebooks where we sketched what we saw in the sky. We met in the crisp autumn night, out past the fields where students played soccer and field hockey, and stood in a semicircle. He pulled out a row of small Questar telescopes, no more than 18 inches long and 6 inches wide, and walked us through the different parts. *Here's how you remove the lens cap. This is the eyepiece. Look through it and you'll see a pair of crosshairs intersecting in the center. That's where you want to put the object you're looking at in the sky. But look. If you put a star at the center of the crosshairs, wait fifteen minutes, then look through the telescope again, the star won't be in the same place. Do you see that? Why is that?*

We talked about how the Earth moves. It rotates—360 degrees in twenty-four hours. Wherever you are right now reading this, stand up and turn around in a circle. That didn't take you very long. But the Earth is much bigger. So it has to turn a distance forty million times larger around before it has done a full rotation. It takes about twenty-four hours to do this. Three hundred sixty degrees in twenty-four hours means that it turns 15 degrees per hour. This is how fast the Sun moves across the sky throughout the day, and why it does.

Fifteen degrees in one hour is a lot. If you hold up your pinky finger at arm's length against the sky, the width of that finger is 1 degree. It wouldn't take long for a star or a planet or even an entire galaxy to drift right out of the little field of view you'd see through the eyepiece of a telescope. So we did something called polar alignment. Mr. Harper showed us how to point the mount of our telescope at the North Star, Polaris. Then we turned on the clock drive. This is an interior motor that keeps up with the Earth's rotation, so that whatever we put in the center of the crosshairs of our eyepiece stayed there. We had learned in class how Polaris looks stationary in the center of pictures showing "star trails"—concentric circles of the light from surrounding stars. We learned that was because Polaris is the star that the North Pole—and the axis of our planet that runs through it—points to. The rest of the celestial sphere over our heads appears to revolve around Polaris as the Earth rotates through its twenty-four-hour day.

My alarm clock beeps loudly in my ear. It's 3:30 a.m. I feel that striking moment within a dream, when I am awake yet not fully aware. I slowly ascend from the immersive deep, craning toward the shore of consciousness. I grasp it, and emerge above water. I collapse, exhausted. The alarm becomes part of the here and now. *That's right. Our class is meeting out past the fields at the observatory.* This needs to be done at 4 a.m. Jupiter is in the sky in the wee hours of the morning in early February.

I pile on three layers—undershirt, turtleneck, sweater—followed by puffy blue parka, hat, mittens, and snow boots. I grab my backpack with my logbook, pen, and a flashlight, and head downstairs and out the front door of our dorm. It is pitch black. It feels like I am the only person awake in the world.

As I trudge through snow-covered fields, about six inches deep, I start to see other dark shapes converging on our shared destination—two large white domes, painted that color to reflect the sunlight during the day and keep the instruments cool inside. We are all headed to the same amphitheater to enjoy the quietest show in the history of public events.

We set up our Questars in a semicircle outside—two people per telescope, so we can help each other out. We polar align our telescopes and are told to find Jupiter in the sky. It isn't hard to spot. In a sea of tiny points of light, a planet easily stands out. Planets don't twinkle. Jupiter's light takes forty-three minutes to get to us, standing in a semicircle in rural New Hampshire. Light from the closest star—Proxima Centauri—takes 4.3 years. Light from the nearest galaxy to us—the Andromeda galaxy—takes 2.4 million years. Once that light gets here—once we see that galaxy through a telescope—who knows what the galaxy looks like in real time, 2.4 million light-years away, where it sits in a spiral of stars in the silence of space? Is it even still there?

The light from stars passes through so much more "stuff"—space and dust and gas and our planet's atmosphere, which tickles and dances in a star's light, making stars twinkle. We call this *scintillation*. I like *twinkling*. It's one of the first songs children learn. *Twinkle, twinkle, little star, how I wonder what you are . . .*

"Is it twinkling?" I ask, in another time and place.

"No," she says, in that other time.

"Then it's not a star. It's a what, honey?"

"A planet!"

The light from planets doesn't have as far to go or as much stuff to pass through. Their light hasn't been as spread out. Planets stare right at us, and we can see them, straight and pure and constant. No twinkling. At least, not from planets in our own solar system. Planets around other stars are another matter entirely. Here, now, it's 1992, and no one knows anything about those other planets yet.

I point my telescope at the tiny pinkish white blob that looks bright and steadfast in the southwest sky. I put it in the center of the crosshairs of my eyepiece. I stare. The first thing I see are thin bands across the planet's disk. These are atmospheric gases and swirling clouds—ammonia, hydrogen sulfide. Ammonia ice. The atmosphere is active. And to the side, down in the lower third of the planet's disk, a large red oval—a giant hurricane.

I watch. I don't want to look away, for fear of this thing disappearing. Is it really out there, a giant ball in space? Then I see something else. Tiny dots. This is what we're here for. I count four of them. I would learn that there are forty-nine others. But these are the biggest. Big enough to have formed themselves into spheres. They appear as dots suspended next to Jupiter—three on one side, one on the other. They are Jupiter's moons—Io, Europa, Ganymede, Callisto.

I sketch what I see in my logbook. I note the time. Mr. Harper passes out hot cocoa. I let the paper cup warm my fingers, and place my nose and mouth above the lip of the cup. The steam wafts up into my face. The smell is rich milk chocolate mixing with the freshness of the crisp air. My nostril hairs begin to relax. The stars burn brightly over my head.

Later, I look through the eyepiece again. The location of the dots has changed slightly. I make notes on position, time. After some more time has passed, another look through the eyepiece. More measurements. We head inside the main observatory building, take off coats, and look at our data in the red light that illuminates the room so as not to ruin our night vision. We sip cocoa as we refine our sketches and check our measured positions and times with others. We will discuss it all in class. But not today. Mr. Harper has canceled the regular class time so that we can go back to bed.

The period of Jupiter's moons—that's what we were there to measure. We needed to know how far each of the moons moved over how long a time. That's all a period is—how long it takes to do something. From a small measurement on that chilly February morning, of a small distance covered in a short amount of time, we got a rate of travel. From the knowledge of the average distance between Jupiter and each of these moons, which we looked up in catalogs, we had all we needed. Io is closest, orbiting around Jupiter in just under two Earth days. Then comes Europa, which orbits in three and a half. Ganymede next, at a little over seven. Then Callisto, the farthest out, orbiting Jupiter in a little less than seventeen days.

That I could measure something in space, just by looking—this was the shattered ceiling of the Earth, ascending up and through the atmosphere into nothing, and everything else. Everything up there was waiting for me to reach out and unlock its secrets. Nothing existed unless I watched and noted it in my logbook. And at the same time, I trusted that, as I laid my head on the pillow to recapture lost hours of sleep, the celestial sphere would continue to revolve around my bedroom in my dorm, within my little corner of New Hampshire, in the frost of a winter hemisphere on planet Earth. This was the tremendous selfishness and audacity that I claimed, alongside my awe and humility in the face of galaxies full of stars, which would eventually lift their veils to reveal orbiting planets. I knew inside that all of this we call the universe was for everyone. At the same time, it only felt tangible—personal—if it was mine, for me alone. I wondered if it must be like that for everyone. I didn't care much, though, if it wasn't.

☾

In my senior year I became a proctor at the observatory. I was given a set of keys to the domes and the main building. I unlocked the doors and set up the telescopes for public viewing nights. When the Moon was out in the winter, people lined up to see its craters and dark basalt seas—called *mare* (pronounced *maw*-ray)—up close, a blanket of snow reflecting the Moon's light back up into the sky and making it seem almost day. When Saturn came into view I trained the 14-inch Celestron reflector telescope on it and encouraged people from town to look through the eyepiece. I watched as their eyes got big and they pulled away to look around to the telescope's mirror. Then they followed the instrument's direction with their naked eyes, up through the rectangular opening in the dome, and into the distant sky, searching for the planet they had just seen up close. Then back to the eyepiece. They couldn't believe that what they were seeing was real. Saturn through a telescope looks exactly like a sticker. The rings—majestic bands circling a planet so tenuous and light that it would float in a bathtub large enough to hold it—are easily visible with relatively low magnification. Some objects, like comets or stars, give only fuzzy details through a telescope. They are too far away or too much like blobs to give you much up close. Saturn is different. All of that ice and dust bound by gravity into hula hoops around the planet. They turn the planet into a queen, with halo wreaths, ever honorable and stalwart, suspended and breathtaking in eggshell hues.

The observatory was my shelter. Whatever happened back across the fields on the main part of campus—bad exam grade, guy I liked not liking me back—I could always go there and open the dome, turn a telescope to something steady and beautiful in the sky, and be home.

The observatory gave back too. I brought the boy I had a major crush on that year there. In a bold move, I'd asked him to the senior dance—our version of the prom—after barely speaking to him for most of the year. He wasn't much of a talker. He frowned a lot. But he was a straight shooter. If he didn't like someone, he didn't pretend to. What you saw was what you got. I liked that. I also liked how aloof he was. I liked the challenge of making him smile, and thought about what that would feel like. I knew if he said yes, I would warm him up. He did. I brought him to the observatory and showed him planets and the large bright Moon through the telescope. Then I kissed his cheek. He didn't know what to do with that.

After I closed up the dome, I found him sitting on a grassy hill outside, his arms folded around his knees. I sat down next to him. I told him it was okay. He turned and put his arms around me, brought his face close to mine under his baseball cap, and guided me back onto the grass.

"You're a strange girl," he whispered low and deep.

"Is that a good thing?" I said.

"Yes," he replied.

I remember how my lips felt warm and curved into his curves and grooves. I remember gentle stubble grazing my cheek. I remember that he asked me if I wanted him to stop. I said no. Not stopping meant we would kiss and kiss and kiss, and that I would have chafed lips for a week afterward. Not stopping meant he would raise my shirt and unhook my bra in one quick gesture, and find my nipple with his mouth and tease it with his tongue, tracing circles around and around. Not stopping meant, when I got home that night, still a virgin, the magic went with me, without any of the worry of what the hell did I do and what could happen. Not stopping meant that I was safe, and still so very, very different from anything I'd ever done in my good-girl life, and reliving that perfect moment in time over and over again, after the play-by-play with my dormmates in my room after curfew. Later that year I would mess it up, grab hold too tight, and we would go to the senior dance at arm's length. After graduation, I would try to cross the chasm, then get drunk and fumble through a conversation to win him back, to no avail. I was always trying to get back to that perfect moment on the hill with no fear and no blame.

For my senior astronomy project I'd focused on astrophotography of the Orion Nebula—a cloud of gas 1,344 light-years away where stars were being born. It's what you'd see if you pointed a telescope, or even a pair of binoculars on a dark night, at the middle star in the "sword" of the constellation Orion the Hunter. In that stellar nursery lie about seven hundred stars in various stages of life. People are used to seeing lots of colored gases in pictures of the Orion Nebula, the result of all of those electrons calming back down after getting excited by the heat from those baby stars. I wanted to get that gas literally out of the picture, so we could see the stars. I don't remember why that mattered to me. I suppose it was because it had never been done before. At least, not that I had seen. Mr. Harper taught me how to develop photos in the darkroom of the observatory, gently dipping the photographic paper in one tub of solution followed by another. I hung the paper on taut strings across the room with clothespins and watched as blank white turned into shapes. Four dots remained in a sea of black night. They were the four brightest stars of the Trapezium, an open cluster of stars in the center of the nebula. No wispy gas showed up. I'd gotten the exposure time right.

Today, these types of images are taken using charge-coupled devices, or CCDs. Everything is electronic. No darkrooms or tubs of solutions needed. Even most amateur astronomers use CCDs. Most observing is done in cozy, heated control rooms, or better yet, from the comfort of your office on the other side of the world, remotely telling the observatory operator the coordinates of right ascension and declination—the latitude and longitude equivalent on the celestial sphere—to plug into the computer to locate your object in space. No manual handling of the telescope, polar aligning, or standing frozen in the cold night air needed. Sounds like a good deal. But all I know is that capturing those stars on grainy film for myself, the old way, made those large distances between us—the stars and me—nonexistent. They weren't trillions of miles away anymore. They were in the palm of my hand.

Jazz

I never intended to do anything else at Exeter besides study astronomy and have a boyfriend. I liked playing Vivaldi's *Four Seasons* on the violin in the chamber orchestra. I even tried out for the first-ever women's varsity softball team and made the cut. But their place was clear in the hierarchy of what I deemed important. These were all extracurricular activities.

One day I auditioned for a play. The theater department was mounting a production of *Steel Magnolias*, and I was coerced by some far more enthusiastic dormmates to go along with them to audition. Every girl on campus would have killed someone for a role in that play. Except me. I couldn't have cared less whether I got cast. Going to an audition seemed like a great distraction from homework, so I went along, then promptly stopped thinking about it.

One afternoon I sat watching the Nicolas Cage and Laura Dern movie *Wild at Heart* on VHS in a packed TV room with my dormmates. It was an extremely racy movie for high school students. For any audience, really. Voodoo and lots and lots of sex. The room was dark, all of us riveted and feeling naughty for watching a movie our parents would never permit us to see in the theaters if we were living at home. Katherine, a lanky blonde senior on the crew team with a kind face, opened the door and peered around.

"Aomawa!" she whispered into the darkness.

"Yeah?" I whispered back.

Her eyes found me. "You got cast!"

She seemed as surprised as I was. Of all the hundreds of girls who'd auditioned for this play, I was one of the six who had been chosen. I had won the lottery. So had Katherine.

Something new started. At first it was logistics. Meeting the director, Mr. Lane, a round, jovial man who could have played Santa Claus. Meeting my other castmates—all white girls from across campus, in their own way resembling the characters I'd seen portrayed in the popular movie based on the play. The girl who would be playing the beautiful, ill-fated daughter, Shelby, actually looked a little bit like Julia Roberts, with her long, chestnut-brown hair and doe eyes. And then there was me, playing Truvy, owner of the hair salon, where the entire play took place. Dolly Parton had played her in the film. What in the hell were they thinking? At the organizational meeting, Mr. Lane told each of us why we had been cast. When he got to me, he told me I was "striking." And that was it. We got to work.

I watched as a small black box theater grew walls and mirrors and spinning chairs and plants and windows and light. A beauty salon was built on the stage. The sink really worked. I could wash hair onstage. Mr. Lane told me to get a manicure at home over winter break, for research. What fun homework!

Something happened to me in the telling of this story of Southern white women's trials and friendship. I wanted it to keep going. Black, white—who played what didn't matter. I was a part of something. It felt different from the sky, where it was always just me and the stars. This was a community, a shared life. We were all working to create something bigger than us that we believed in. When it was ready, we would open its door and let the world in. We would tell the entire school and the town a story, and we would do it well. We would make them cry, because they would believe it. I felt the power in that, in having a secret shared with my cast, and knowing that at a predetermined time, the secret would be revealed and it would blow everybody's heads off.

Something about this was familiar. It was foggy, but I started to remember bits: I am in the third grade at Torrey Pines Elementary School, back in La Jolla. I am in a play. The play is called *The Rabbits Who Changed Their Minds*. I am not just in it. I am the Rabbit Queen. I am commanding, foreboding, strong. I take charge of the stage. Everyone is impressed. The freckle-faced popular girl in my class, with feathery blonde hair clipped two sides back and white Reeboks, even stops terrorizing me for a month afterward. I make her think twice. If I can talk like that onstage . . .

The scenery changes again. I am in the car late at night with my family on the way to Ottawa, Ontario, Canada, no road signs for motels in sight, and my grandmother starts clapping her hands in the passenger seat to keep my mother awake. This is her nightmare scenario come true. We are one long eyeblink from running off the road or into oncoming traffic.

So I start at the beginning, reciting in its entirety the musical *Shout Up a Morning*, which I recall completely from having understudied a role in the play a few months earlier, in the fifth grade. I start in act 1, scene 1—John Henry coming up the railroad tracks—and I keep going until the curtain closes on the final act. I had sung the very grown-up song "Blues in the Night" for my audition for the professional La Jolla Playhouse production. I hadn't gotten the part, but as an understudy I went to every performance over an entire month.

That night I recited all of the characters' lines, sang all of their songs. I kept my mom and grandmother awake and entertained. They had both seen the musical many times. Someone had to go with me each night. But through my voice and interpretation, it was new and different and special. I loved that feeling of being new and different and special.

The week before *Steel Magnolias* opened at Exeter, I got sick with a terrible cold. My mom, grandmother, and brother drove up for the play. They whisked me to a nearby hotel and poured hot tea with lemon and honey down my throat. There were no understudies for this high school play. There was no rescheduling. We had rehearsed all term long for four shows over one weekend. I would not miss any of them.

On opening night, my good friends Yvonne and Phil sat in the front row, clutching hands as they waited for my first line, spoken as Annelle finished doing my hair for her job interview. I opened my mouth, and a gravelly grind like steel scraping concrete as brakes ran dry came out. I kept going. There was nothing else to do. Truvy had a cold. She still had to work.

Even with my unfortunate debut, I was hooked. Telling a story, making an audience feel something they wouldn't have felt otherwise, the camaraderie of a cast, rehearsing day in and day out, performing onstage and then leaving that world to return to our own lives, only to do it again the next night, and again. I couldn't get enough of it.

As the daughter of two musicians, performing was a fact of life. My parents met at Antioch College in Yellow Springs, Ohio. Their band was called the Pyramids. They toured Africa and Europe playing jazz, Afro-jazz, and other African-influenced styles of music. My mother played the flute, my father the saxophone. They learned dozens of other instruments.

I, naturally, played a musical instrument too. First it was the piano, when I was six. Then my teacher went back to Switzerland, and I lost interest in the piano. One Christmas at my stepfather John's family home in Ottawa, Canada, I wandered down to the basement and saw a violin lying abandoned in its case in the corner with one string on the fingerboard. I asked my stepfather if I could play it. I played in orchestras and chamber music groups. I even played backup violin for the R&B singer Brian McKnight.

My father wanted me to try playing jazz. I recoiled. Jazz is all about improvisation. There is a generally understood main theme that everyone starts with. Once that melody is established, the song cycles through each performer, giving the player the chance to showcase their instrument and their ability to improvise, playing around the melody, underneath it, above it, countering it. But everyone always comes back to and stays true to the key and the main theme of the piece.

This terrified me. I was comfortable with a piece of music that was written out for me in its entirety. Classical music was like this. I knew how to read music, so as long as I knew how to play the notes I read on the page, I knew how the song would end, and I knew I could play it. With jazz, knowing that at some point I was going to have to "riff"—wing it for measures on end—was overwhelming. I felt like I didn't know my instrument well enough to do that. I didn't want to take the risk that I'd play out of key and screw up the whole piece.

I didn't realize it as a child when I formed my disdain for jazz, but this early protest said something about who I was. I was someone who was completely uncomfortable with uncertainty. In fact, I was petrified of it. But this is what life is—uncertainty, day in and day out. As much as I pretend to know what's going to happen next, and plan my days and weeks as if I do, rarely does anything go the way I plan. You know what I'm talking about. I didn't like the prospect of changing my career plan midway through high school. I'd written the notes for how this piece of music was supposed to go back in the seventh grade. In ink. Still, I couldn't ignore what I was discovering about myself either. I loved acting. And, on top of it all, a career in acting was about as uncertain as one could get.

One play led to another, and soon I was on the board of the student-run club Dramat—planning the productions for each year and acting in half of them. I kept my role as proctor at the observatory, though. Here, in this small world of a private boarding school community, I did both, and no one seemed to question that choice. So I didn't either.

That's how it was and has always been for me for as long as I can remember. The stars, the night sky, the stage, and the screen. It never occurred to me until much, much later how unusual it was to love two things that appeared to be so different. Most people seemed to gravitate to one thing. Business or architecture. Engineering or fashion design. Science or art. Is that really true, or a story I'd been told—that one must choose one straight road and stay on it? No bathroom stops, no diversions.

At Exeter the reality set in that the arts were probably my most natural inclination. My steadfast goal of going to space hung suspended in the sky. Improvisation was going to get me after all.

High Flight

In my junior year at Exeter I had to choose a topic for the dreaded History 333 paper, a term paper all juniors—or "Uppers" as we were called there—feared. This paper was a serious piece of writing. We had to include a bibliography and footnotes. Each of us carried around a box full of index cards with each major point we had cited from a reference. We were learning how to form a thesis and back it up with supporting documentation without plagiarizing. We were being taught how to tell a story from U.S. history like scholars.

I chose the 1986 space shuttle *Challenger* disaster as my paper topic. The year the *Challenger* blew up, I wasn't interested in going to space yet. I wanted a sandy-brown-haired boy named Joshua in my fifth grade class to like me back. I loved my best friend, Katie, and rambled around the playground with her during recess. Our parents took us to the mall to see movies together. I had lived in Victoria, British Columbia, Canada, for a year with my stepfather while my mother finished her PhD. Finally I was back. The blonde, freckled girl had stopped bullying me for good. I had been to another country. I was cosmopolitan. I'd permed my hair. I wore pastel clothes. I loved Cyndi Lauper and Olivia Newton-John and

Wham! and Michael Jackson and Men at Work. I belted out lyrics to their songs out on the fields with Katie while we played tetherball.

One morning I walked into class and noticed that Mrs. Biondo's face looked red, like she'd been crying. I sat at my table opposite several of my classmates and asked them what had happened. A nerdy boy at the end of the table with brown wavy hair, glasses, and lots of freckles, wearing a Boston Bruins hockey jersey, said, "Haven't you heard? The space shuttle *Challenger* blew up."

The teacher on board, Christa McAuliffe, who was to be the first teacher in space, carried the hopes of every other teacher in America with her. Unlike the other astronauts, who'd trained for years as fighter pilots and PhD scientists, Christa was a schoolteacher and a mom. She was proof that an Everyman—who in this case was an Everywoman—could have the most impressive job in existence. A regular person could ride atop a rocket and blast off into space! But she didn't.

Two years later, when I saw the movie that made me want to be an astronaut, I forgot about Christa McAuliffe and the *Challenger* disaster. I wanted that life, risks and all. My mother, friends—anyone I told about my career aspirations—would bring up the *Challenger* explosion as a reason why I was absolutely out-of-my-mind crazy to want to be an astronaut. But it didn't faze me. I was patriotic, and I told them how honored I would be to give my life in that way for my country.

I wasn't alive when John F. Kennedy was assassinated. But I remembered exactly where I was and what I was doing when the *Challenger* exploded. It occurred to me that this thing that had happened back in the fifth grade, that I was alive to witness (if not firsthand), was a part of United States history. I wanted to write my paper on it. It felt closer to me than the many pieces of history that occurred outside my own personal timeline.

After the initial horror of that tragedy subsided, what remained was a technical detective story. Why had the shuttle exploded, and just seventy-three seconds after liftoff? Twenty-four prior shuttle flights had returned their crews and payloads safely. This particular shuttle—*Challenger*—had flown nine times before without failure. What specifically had gone wrong here?

I spent hours combing through encyclopedias and authentic sources. I read the report of the Rogers Commission that President Ronald Reagan had appointed to determine the cause of the disaster. Neil Armstrong, the first man to set foot on the Moon, and Sally Ride, the first American woman to fly in space, were on the committee. I read articles on microfiche, staring into a small booth at tiny illuminated text, looking for something definitive.

I believed in the safety of construction. It always baffled me to hear about bridges collapsing or faulty elements in cars that required entire fleets to be recalled. I suppose I expected engineering to be perfect, when nothing is. When I learned that the problem was in the O-rings, it didn't seem possible. How could they—whoever *they* was—not have known this could happen? How could they not have known enough to stop it from happening?

It sounded like some kind of conspiracy theory movie. That something that important could slip through the cracks. Or perhaps just be ignored because the potential impact was inconceivable. I suppose the unthinkable is called that for a reason. No one wants to think that terrible things could happen on their watch. It's too horrible a prospect to entertain. And perhaps that's why not enough was entertained prior to launching seven people into space on a rocket. I prefer to think of hubris as a reason rather than impatience or blatant disregard. Who knows for sure about those "communications" and how they "faltered"? Someone somewhere who will never admit the mistake. Who would? We're humans. I don't like to admit mine.

Ronald Reagan included portions from John Gillespie Magee Jr.'s poem "High Flight" in his televised eulogy for all seven of the astronauts. He said that they had "'slipped the surly bonds of earth' to 'touch the face of God.'"

Senior year at Exeter I read the full poem during a meditation I wrote on my future dreams and goals of becoming an astronaut. I called the meditation "High Flight," and at the beginning I read the poem. The piece was about the journey I took to discovering that I wanted to be an astronaut. I read it to a crowded room in Phillips Church, as other seniors did every week during that spring before graduation. There was actually a point in the piece when I said out loud, "I mean, I'm actually going to do this!" People cheered, and I was very proud of myself.

In the years since writing my 333 paper, I've thought a lot about *Challenger*. I no longer disregard the peril that accompanies a career riding atop a rocket full of pressurized burning gas. But that isn't what I think about most often when I think of *Challenger*. I think of the crew. Every January 28, I remember their faces—not only Christa McAuliffe's, but Judith Resnik's, Ron McNair's, Gregory Jarvis's, Ellison Onizuka's, Francis "Dick" Scobee's, and Michael Smith's. I look at the promotional photo NASA took of them all in their blue flight suits. In the picture, Mike, Dick, and Ron sit in front, their hands gently cradling the helmets on the table in front of them. On Mike's right there is a miniature model of the space shuttle on the table, pointing up and poised to lift off at any moment. In the back row stand Ellison, Christa, Greg, and Judy. They are holding their helmets in the crooks of their left arms—all except Ellison, who balances his on his right knee, which is propped up on the bench that Mike sits on, beside Mike's right hip. Ellison's right wrist and the right side of his hand, turned slightly outward, rest on top of his helmet. The United States flag towers behind him, outstretched to extend to just behind Christa's right shoulder. I wonder if some unseen person is crouched behind her, holding it there.

In this NASA promotional photo, everyone looks so—normal. Their smiles are genuine. They look happy, excited. To have this job. To have been chosen, from thousands of applicants, to sit and stand there and pose for NASA. To go to space. Judy and Christa both have curly, big hair, the way most white women would have worn their hair in the '80s. Judy's skin color is a golden bronze compared to Christa's, which looks like peaches and cream. I see blush on Judy's cheeks. And eye shadow. She was an electrical engineer, a software engineer, a biomedical engineer, and a pilot. She also cared about her looks. They could all have been posing for a photo at a barbecue.

I wonder about their families. What had their lives been like over the years? Did the million-dollar settlements from the O-ring company and the government make their lives easier? Judy's ex-husband of twelve years was the lawyer who represented her father in the civil lawsuit. She had no children. Dick had a wife and two adult children. Mike had a wife and three children. Ellison had a wife and two small kids. So did Ron. Greg had a wife. Christa had a husband and two children, Scott and Caroline. They were nine and six when she died. The older, Scott, would be forty-six now, two years younger than me. What had their lives been like? Did they think about their mother every day? Did they even remember her? I start to make it about me. If I had kids and then died when they were young, would they remember me?

I look up Christa online and see a picture I've never seen before. It is another NASA promo photo. But this one is of Christa with a woman sitting a little lower and to her right, with the American flag behind her. The caption reads that the woman's name is Barbara Morgan. She was the backup teacher in space. Christa's backup. She has shoulder-length, straight, dark-brown hair, one side back, and a side part above the left corner of her forehead. She looks young, tired, and happy. This woman, whom most people have never heard of, came so close to being famous for her death. Instead, she lived. I read on. She carried on the legacy of the Teacher in Space project. It didn't die with Christa. Barbara spoke widely, was an educational consultant, designed curricula, and even served on the National Science Foundation's federal task force for women and minorities in science and engineering. I probably owed one of my later NSF grants to some of the work she did while on this task force. She continued to teach second, third, and fourth grade in McCall, Idaho. Twelve years after the *Challenger* disaster, Barbara officially became an astronaut. She logged twelve days, seventeen hours, and fifty-three minutes in space before she retired from NASA in 2008.

At Exeter I could do it all—acting, astronomy, softball, orchestra. I was a renaissance woman, and I was rewarded for it. I won the drama prize and the astronomy prize. I won the French prize. And the prize for the study of twentieth-century African history, for which I received a book about the Golden Age in Europe.

At graduation I wore long, flowing white pants with a crocheted white sleeveless top in daisy patterns, and killer white platform sandals. My mother had taken me shopping for the outfit back home in Amherst. We'd gone to the hippest clothing store in town—Zanna—where a small woman with close-cropped chocolate-brown hair and thick, bushy eyebrows presided over racks of carefully selected dresses, blouses, and slacks.

All the senior girls wore white to graduation at Exeter. The boys wore suits. I don't know why we wore white. Women had only been allowed to study at Exeter for less than twenty-five years at the time. Maybe the only way we could stay was to seem like pure and untouched virgins who would never tempt the boys from their studies and big plans. Maybe it was a choice the first female Exeter graduates made themselves, and it stuck. Perhaps they were inspired by pictures of women wearing white while marching in the suffrage parades of the early twentieth century. White women marched in the front of these processions; when Black women were permitted to participate, they were expected to march separately. Perhaps the early Exeter graduates were also inspired by the Black women and children who wore white at the July 28, 1917, Silent Protest Parade on Fifth Avenue in New York City, organized by the NAACP. Ten thousand African Americans marched on that day in opposition to the lynchings and other anti-Black violence and police brutality that had escalated over the prior year into a twenty-four-hour assault on Black people in East St. Louis, Illinois, earlier that month. They marched in memory of their "butchered dead." They marched to show the world that Black lives mattered. They marched in support of a better world and better lives for their children.

When the principal called my name as the winner of the Perry cup I was stunned. I don't think I'd ever been so surprised by anything in my life. I physically jumped about half an inch out of my chair. The silver cup was awarded to a graduating senior for demonstrating the qualities of outstanding leadership and school spirit. It was named for Edmund Perry, an African American boy who'd come to Exeter to pursue his American dream of higher education in the early 1980s. He'd graduated, then gone home to Harlem, New York, for summer vacation, only to be killed in the streets by a white police officer. The policeman claimed that Perry and his brother had tried to mug him in a park in Manhattan. When the press found out that Perry had gone to Exeter, the killing made big headlines. Books and made-for-television movies followed. Perry's mother died six years later of a heart attack at the age of forty-four.

My math instructor, Ms. Kay, a short, red-haired tomboy who made math look simple and elegant, and was also my dorm resident mom, said later that she had been looking right at me when the principal read the qualities embodied by the recipient of the Perry cup, before announcing the name. Ms. Kay said she couldn't think of anyone else more deserving of the cup.

I'd loved Exeter with all of my heart. I was truly happy being there. There, I felt no limits, no ceiling to what I could achieve. Despite the push-pull of race, color, finances, and boys, Exeter had given me a world to call my own amid it all, like Jimmy Santiago Baca wrote about in *A Place to Stand*. I could explore who I was, or at the very least what I wanted to do, and no one tried to put me in a box.

I still think of the Magee poem, and in particular the line "the delirious burning blue," especially when I fly. Thirty thousand feet above the surface of the Earth, heading west, I follow the curvature of the Earth, and sunsets last forever as I chase the dying light burning rosy red along the grazing edge of the horizon. I watch the Sun's final embers dance from the rooftop of the world, hoping to catch up with the Sun if only the plane moves fast enough. Up there, above even clouds, there is only silence, and great swaths of blue expanding for miles. This is why I used to request the window seat as a child, rather than the aisle as I do now.

My life felt like that view. A great open space, a clearing, with everything I'd seen and experienced down below—the lush green squares of Missouri after a flood, the bone-brittle browns of a drought-laden California. It passed below me, and it was my turn now to look above. Now there was only possibility.

CHOOSING

Physics

When it came time to apply to colleges, it became clear that after having the freedom to explore many fields and interests at Exeter, I would finally have to make a choice. If it was the wrong one, then I could make another one. But I had to move forward. I held MIT close to me like an old friend. Really, I had chosen when I was twelve. It was the plan. As much as I tried to open my arms to other possibilities, and good ones too—like Stanford, where I would have gotten a liberal arts education, along with excellent technical science or engineering training—I couldn't make the sale on anywhere else. My father, still in California after all these years, was disappointed not to have me in closer proximity to him for college. But how could he chastise his daughter for choosing to go to MIT? It was impossible.

I was admitted early decision to MIT. I remember in the interview that the admissions officer asked me how long it would take a spacecraft to travel to Mars. Maybe he thought I wouldn't know, or would falter or fold on the question. But I knew. The Mars Global Surveyor mission, whose launch was on the horizon, would take about ten months to make the trip. I wonder if the admissions officer checked my answer after I left. If he did, he must have seen that I was right.

I couldn't believe that I was here, even after all of my careful planning. Suddenly here it was, a dream realized, to study at the most prestigious institution for science and engineering in the world. The large, manicured lawns, the long concrete columns of 77 Massachusetts Avenue, steps leading up to two iron doors. Inside, a long hallway called the Infinite Corridor. It went on and on for 823 feet, branching off into hallways leading to the departments of civil and environmental engineering, urban studies and planning, aeronautics and astronautics, mechanical engineering, physics. On several days during the year, the Sun sets in alignment with this corridor, filling its entire length with golden light. Then, outside, an open space, pointing the way to other buildings, most of which were built with gray, sturdy concrete. Biology and chemistry labs. The medical center. The tallest building on campus was the Green Building. It wasn't green. It was gray concrete too, windows in rows all the way up, like packing foam. It housed the Department of Earth, Atmospheric, and Planetary Sciences—EAPS, for short. At the top was a weather station.

We came from everywhere to fill the lecture halls of this old, illustrious school, on the outskirts of the Ivy League because MIT wanted it that way. Those techies were in a class all by themselves. They didn't engage in petty rivalries like Harvard and Yale. They pranked football games. MIT students put a fire truck on top of the Great Dome, one of the main buildings on campus. They stole an antique cannon from the lawn of that other technical school, Caltech, and moved it to their own campus. They made Caltech travel across the country to retrieve it. MIT students were—true to the school's mascot, the beaver—resourceful. They knew how to do stuff.

Ten thousand students studied at MIT—five thousand undergrad, five thousand grad. There were fraternities and sororities. There was even a Division III football team. MIT tried to reflect the full college experience.

PHYSICS

During the early formation of the Earth 4.5 billion years ago, the planet was new, fresh, and molten. This early stage is called the Hadean Eon, because if you could have visited the Earth back then you would have seen a hellish world of fire and lava as the planet continued to settle. Rocks fell from the sky—pieces left over from the formation of the solar system, flung through space by the gravitational influences of far more massive bodies, like Jupiter. This period of time, full of constant impact into Earth's surface, is called the Late Heavy Bombardment. I think of it affectionately as the physical embodiment of my freshman year at MIT. As everyone said, the first year is like drinking from a fire hose. Sometimes getting what you want is more than you can handle.

I decided once and for all that I would no longer be an outsider around other African American students as I had been in high school. I didn't want to do the same dance that I'd done at Exeter, worrying every day in the dining hall about whether sitting with the people I felt most comfortable with meant turning my back on my heritage. I wondered if there was a way to feel comfortable and accepted around other Black students, even if I hadn't grown up in Black neighborhoods or gone to Black schools. My desire to belong strained and stretched against my desire to say *fuck it*.

I ended up doing a little bit of both. After nearly accepting an offer to join a co-ed fraternity simply because it was full of students with accents from all over the world who smoked—nothing like the typical crowd I imagined occupying fraternity or sorority houses—I reached out to two Black women I'd met during orientation. I asked them if they wanted to room together. They said yes. They were planning to live in the all-Black dorm on campus. So I planted myself in a crowded double room with them and got ready for classes.

All courses at MIT are known by their numbers. All physics courses start with the number 8. All EAPS courses start with the number 12. Instead of saying you were a physics or an EAPS major, you said you were

"Course 8" or "Course 12." Everyone instantly knew what subject the numbers mapped to. We used any excuse we could get to code in our heads.

The first physics course I took was 8.01 (pronounced "eight-oh-one"; you don't mention the decimal point). Classical mechanics. Pullies, pendulums, and projectiles, with force and velocity components in all sorts of directions. I was failing. My professor pulled me aside when he found out about a weekend field trip that my advisor was organizing for his advisees at his ski cabin. We were going to hang out and drink hot cocoa and play in the snow. The 8.01 midterm was the following Monday.

My professor told me, "If you go on that field trip, you will fail this midterm." He wasn't threatening me. He simply knew what I hadn't yet discovered about myself: I could not simultaneously give two things 100 percent effort. If I split my focus during a time away from something I was already struggling with, holding on by the thinnest layer of the edge of my fingernails, I would lose my hold completely.

I did not go on that cabin trip. Instead, I lived in the library the entire weekend, only leaving to shower and sleep a few hours before returning. Libraries at MIT were open all night. I ended up scoring fifteen points above the class average.

It wasn't as high as either of my roommates' scores, though. Shellyann was a slender Jamaican woman with a warm smile and a butter-smooth lilt in her accent who met the guy who would become her husband the first week of freshman year. She would eventually become a doctor and the head of her department at a major hospital in Atlanta, Georgia. In college she had the neatest handwriting I had ever seen. Even with the love of her life already found, rarely apart from him, she scored straight As on physics, chemistry, biology, and math exams.

Jonna was from Florida by way of Jamaica, but Jamaica was in her veins. She was another genius, who loved to eat food from Bennigan's.

When she ate her favorite chicken dish, she would drop her head toward her chest in rapt ecstasy and dance from the waist up, wiggling and jiggling in her seat. She had the extraordinary gift of being able to study, book open, in the middle of chaos. Music blaring—either Lauryn Hill or Shabba Ranks or A Tribe Called Quest—Jonna would absorb what was in a textbook or her notes amid it all.

Jonna, Shelly, and I were a triple threat. There we were, three Black women—two by way of Jamaica, and me. We walked through campus tall, slim, our smooth, sleek skin in shades of caramel and honey. Most of the time we stood out from the gray stone and the seas of milky faces. And because we were together, that felt all right. We weren't alone. And the secret we carried with us, that may have surprised most of the white and Asian students there, was that we were just as geeky as they were. And we could prove it.

Everyone, mostly boys, hung out in our crowded double, room number 312, in New House 2. We did problem sets together. We had a physical constants-off, where we battled to see who could remember the most physical constants by heart.

"Are we doing CGS units or MKS units?"

"CGS!"

"Planck's constant!"

"6.6261×10^{-27} cm^2 g s^{-1}!"

"Boltzmann's constant!"

"1.3807×10^{-16} cm^2 g s^{-2} K^{-1}!"

"Mass of an electron!"

"9.1094×10^{-28} g!"

Late nights turned into early mornings, and we rolled into 8.01 lecture with bagels and orange juice. We were a row of Black faces in the back in puffy parkas, half asleep, taking notes. No one was later than Jonna though, who strolled in fifteen minutes after lecture started with her

Tropicana juice box. She aced every exam without much more than a look over the notes. For Jonna, there was no stereotype threat. She transcended it.

For me, my head in three divergent clouds, it was a different story. My good grades were hard-won. My bad grades were that much more devastating, because I worked so hard and still got them.

I am thinking of statistical mechanics. I never could understand the partition function. I remember moving closer and closer to the front of the lecture hall, hoping that everything would start to make more sense by mere proximity to the professor. She was a large blonde woman. A female physics professor at MIT! If I couldn't learn it from her, who could I learn it from?

I was in danger of failing the class. So I dropped it. I decided that maybe it was the style of teaching that was the problem. Lectures. No engagement. After three years at Exeter, where classes of no more than fourteen students sat around large oval wooden tables with the instructors sitting next to us, being in a lecture hall full of hundreds of nameless students was a big change.

I went to talk to my new advisor, Jim Elliot. A short man with glasses and a square face, Jim was famous for discovering that the planet Uranus had rings, like Saturn. These days he liked to go up in airplanes above most of the water vapor in Earth's atmosphere and take infrared observations of Pluto when it passed in front of distant stars. This allowed him to learn something more about Pluto's atmosphere, which he had detected for the first time in 1988 using this same technique.

Jim's office was in the EAPS department, in that big, tall "green" building. I had decided after several physics courses not to major in physics. Everyone in EAPS seemed to be having so much more fun. They were climbing mountains and observing with telescopes and hiking out to rock formations to collect samples. That was much more my style than sitting in a physics lab all day.

EAPS still required enough physics to make me uncomfortable, including quantum physics and stat mech. Jim and I decided that I should try taking stat mech at nearby Wellesley College, a private all-women's college in the Boston suburb. Classes were smaller there, more discussion-led. I rode the inter-campus bus from MIT to Wellesley. I walked into a room filled with women—albeit white women—and an Indian American female professor. She spoke so clearly and casually about thermodynamics, equations of state, the basics of probability. She described the partition function as a way of cataloging a system in thermodynamic equilibrium, where neither matter nor energy flows into or out of that system. She wrote neat equations on the board and invited women in the class up to solve them. She told us that the saying "heat rises" isn't technically accurate. Rather, hotter air rises not because it's hot, but because the air molecules are moving around so much that the air itself is less dense than cooler air. It's less dense air that rises. I never forgot that. But at the time, as with every single other thing she said and the other students did, none of it made any sense to me at all. I got a C in the class.

I couldn't blame MIT for my bad grades. I went back there and got a D in quantum physics.

☽

I walked across the Mass Ave. bridge to Boston, escaping problem sets and romantic failures in search of the perfect single wineglass and bottle of wine to drown my sorrows. As I walked, MIT was there under my feet. The bridge had been measured in "smoots"—the unit of length described by the 5-foot, 7-inch height of a pledge named Oliver Smoot who had lain down repeatedly, head to foot, across its entire length as a fraternity prank. The bridge's length had been measured at 364.4 smoots, or 2,035 feet. This fucking school.

Clearly distracted, I couldn't be anywhere else or anything else for anyone else. I remember being on the other end of a phone in a public phone booth in one of the main hallways on campus, with Jonna's and Shelly's ears sharing the receiver with mine, while my grandmother told us that Mom's baby had died. She'd gone into labor the night before, thinking all was fine. The baby had been stillborn. They knew before she delivered. My mom regretted watching a violent prison show that night, thinking the energy from that story had traveled in utero. She told the doctors to give her all the drugs then. She didn't have to care anymore. They named him Kevin. My new stepfather, Kirk, had a thing for *K* names. His brothers were Keith and Kenneth. They would eventually try again and produce my sister, Kayla, two years later, making it all worth the hellfire they'd walked through. But now, here, there was only the present moment full of a despair that I couldn't entirely comprehend. I remember being sad, then feeling guilty. I'd been jealous of Kevin. I'd wanted more care packages and visits from my mom. Now this had happened. I went home for the funeral. Then I returned to school and dove back into my lonely life.

If only I could see myself now, loneliness and all. I thought everything hinged on a result—the right one. I was so happy, fresh-faced, to be at MIT. Even with chicken pox, the last pock crusted over and jagged, ready to fall from the left side of my neck, I was happy to be going. I remember the outfit I wore to orientation. Cutoff jean shorts with the ends rolled up. Creamy white long-sleeve top with a scoop neck and laces up the middle. A mint-green-and-white hooded vest over the shirt. My hair was long. Leaping from my mother's car, bounding over to Kresge Auditorium, the building covered in glass. It is one-eighth of a sphere, with supports in only three places. Everything at MIT holds mathematical significance. Last night I read a line from a poem by Audre Lorde. I fell in love with it: "no reckoning allowed save the marvelous arithmetics of distance."

If only I could have been there, myself as I am now, in the car on the street somewhere, at a stoplight. Pulled over in the red loading zone. I would have called to my younger self. I would have called her over to my car. I would have whispered to her, if she had stopped to listen, if she had believed me, that I was there to help, to save her from suffering. I would have whispered, "Don't forget how whole and complete you are, holding it all, including the poetry." I would have maybe said the line from Audre Lorde to her. Maybe she would have listened, because it had the word *arithmetics* in it. Maybe I would have also shared one from Denise Levertov, the one about the bell ringing out, metallic, and how my whole body sang out what it knew—"I can."

If only I could see her now, I would gently stroke her hand as I said it, give it a firm squeeze, and tell her to listen to the rain when it comes. Because she will suffer, no matter what I say. She will stumble. There will be problem sets and chalkboards in countless lecture halls throughout this campus and beyond that are covered with equations she will forget or never understand how to trace her way back to, because the instructor—Professor Henley, or Professor Wójcik, or Professor Benchley—leaves her in the dust. Outside, the rain is coming down in sheets. We need it. It is beating down. I would tell her about what's to come, and how she doesn't need to worry about what it all means. One day soon, she will get it. She will understand why conservation of angular momentum is so important. It's why when ice skaters pull their arms in toward their bodies, they speed up. It's how our solar system was born, a cloud of gas and dust becoming denser and contracting and heating and spinning up. Out of that cloud emerged the Sun, our star, a ball of hot plasma, hotter at the center, cooling toward the outer shell of the sphere. Then dust and gas began to cool, solids forming according to their chemical structure, the rocky and iron planets like Earth and Mars forming closer to the star because they could stand the heat, the gas giant planets like Jupiter and Saturn forming farther out, where gas could stick around without burning away. She will understand later why all of this physics and the need for mass integrals are so important. You need to add up all of the mass in a star to know how much force it exerts, how it keeps the planets orbiting around it. Once she makes the connection between this work and what she loves, it will make sense. Except, as she turns away to go find her roommates at MIT, I say one more thing. *Inside the rain on the rooftops lies the poetry of the universe too. Don't forget that.*

I returned to MIT for a conference recently. In the morning, I took the Halls Shuttle to Harvard Square. My favorite part of the twenty-minute journey was when we crossed the bridge. The sky opened up, and the Charles River sparkled. The sailboats waved their white sails, tacking. I could see the gilded buildings on MIT's campus. The place I called home for four years of my life, twenty-something years ago now, when I was the age of the students I teach today. When I was here as an undergrad, swimming in a sea of fear and anxiety and desperate to water the tiny seed of a dream of spaceflight, they were not yet born.

Research

The last formal physics course behind me, I was finally free to start taking astronomy and planetary science courses, and to dip my toe into research. If they'd let me.

January was a free month. There were no formal classes. Students could go home or stay and take unconventional classes like basket weaving and pilot's license prep. I couldn't wait to do the MIT Astronomy Field Camp. My advisor, Jim, ran it. Every January he took a small group of students on a plane from Cambridge, Massachusetts, to Flagstaff, Arizona, to do research for three weeks at Lowell Observatory with staff astronomers. I was the first freshman to participate.

Deidre Hunter was a short, soft-spoken woman with large, tinted glasses and straight, dirty-blonde hair. She was the first person to teach me how to do real astronomy research—which meant asking a question about some astronomical phenomenon and then taking the successive steps necessary to answer that question.

Deidre loved her cat, and she loved galaxies. And I learned all about one of them from her. I learned that galaxies—collections of stars—come in all shapes. Some are spirals, like our own Milky Way galaxy. I always try to remember to tell my classes that we have no actual pictures of our galaxy; only artist impressions. The simple reason for this is that no satellites or other spacecraft have ever escaped the boundaries of our galaxy to be able to turn around and snap a photo of it. The farthest-traveling spacecraft—*Voyager 1*, launched in 1977—only recently passed the boundaries of our solar system. So how do we know our galaxy has a spiral shape? We have an idea from looking at the bright strip of the Milky Way that you can see when you go to the desert or some other dark spot on a clear night. We're looking toward the center of the galaxy when we see that strip. This is what we'd expect with a spiral galaxy. It isn't that the other stars outside of that strip in the sky aren't part of the Milky Way. They just aren't part of its central, densest population of stars. The amount of gas and dust in the galaxy, and the velocities and colors of its stars, also are like those of other spiral galaxies we see in pictures.

There are also galaxies that look like footballs, called elliptical galaxies. And there are galaxies that don't have any set morphology (a fancy word for "shape"). These are called irregular galaxies. And Deidre Hunter knew all about them.

I remember the library in Lowell, a cozy, carpeted room filled wall to wall with astronomy textbooks and magazines and volumes of peer-reviewed articles in *The Astrophysical Journal*, *The Astronomical Journal*, *Monthly Notices of the Royal Astronomical Society*. Deidre showed me the first astronomy journal articles I would ever read, full of images of irregular galaxies. Each page was divided into two columns of dense text, often interrupted by an equation. She told me there was a critical density that gas had to have to be able to form stars. There were theories about how easy it was for spiral galaxies to attain this critical density, compared with irregulars. She showed me a small patch in an irregular galaxy called IC 1613. It was a circular shell of ionized gas. Remember, that's gas that's been heated to the point where electrons excite themselves past the overtired state, then crash, releasing that excess energy as light. There were several bright stars in that region, and a big question was whether those stars that were visible in the gas shell actually possessed the energy required to keep ionizing that shell. We were going to figure out whether they did.

Deidre taught me how to use a computer program to identify stars in the image of a galaxy and calculate the amount of light, and therefore energy, emitted by those stars. This process is called photometry, which literally means "measuring light." I sat in the computer room with my astronomy software package manual, in a thick sweater with my long, curly hair around my shoulders and a scrunchie on my wrist, clicking on sources in the image with my computer mouse. I got up from my chair and went to her office on the second floor, where there was thick carpet and a large wooden desk. I asked her questions over and over again. She was patient and never talked down to me. There was always more to learn, more nuance to understand about this study.

There was a star that I'd never heard of before sitting in the circular shell of gas. It was called a Wolf-Rayet star. This type of star was hot—really hot—with surface temperatures of at least 30,000 Kelvin. The Sun is 6,000 K. The difference between Kelvin and Celsius or Fahrenheit isn't important at these high temperatures. Hot is hot.

I learned that Wolf-Rayet stars were at a different stage of their lives than the Sun. Stars spend most of their lives chugging away, making helium out of the abundant hydrogen that they're filled with at birth. Well, Wolf-Rayet stars are massive stars. And when it comes to stars, mass determines everything about a star's future. Massive stars burn through their hydrogen quicker than the Sun ever could. Wolf-Rayet stars have blown off their outer layers of hydrogen and have moved on to making other elements on the periodic table—not just helium, but nitrogen and carbon too.

I learned that for stars, one can be enough. This single Wolf-Rayet star alone could have possessed enough energy to ionize the gas shell in this region of IC 1613. We had our answer. I felt like I had helped to solve a mystery, without touching a single clue. We had used light and our eyes and a little bit of math. We couldn't expect a confession from our culprit, that Wolf-Rayet star. But we didn't need one. Our conclusion was by no means irrefutable. Scientists rarely concluded with 100 percent confidence. But the evidence was compelling enough to write a paper.

I remember that I'd memorized the end-of-field-camp talk I had given. It probably sounded memorized too. But it didn't matter. Everyone—the sea of older white male astronomers, sprinkled with a few older white females besides Deidre—was impressed. At the time, I thought they were impressed because of my age, as I was the first freshman to do research there. I was likely also one of the few Black *anything* to do research there.

I went back elated and relieved to the "Chalet"—the house I lived in with three other MIT students during the field camp. Joannah had a British accent and pale skin. She would eventually marry an MIT classmate, go on to defend her PhD while eight months pregnant, and work for the Multiple Mirror Telescope in Arizona as she balanced parenting a teenager. We shared an irreverence for pompous scientists and professors, and a keen awareness of just how steep the grade was on the hill we were climbing to become female astronomers.

I don't remember the names of the two boys. They were shaggy and seasoned too, and they had weathered many problem sets and the dreaded physics "junior lab"—a yearlong laboratory course that required late hours and the toting around of an orange toolbox full of electronic circuits. I would later dodge junior lab by choosing Course 12—EAPS—as my major.

Joannah, the two boys, and I were now free to lounge in the Chalet living room, walk down Mars Hill and into downtown Flagstaff for an evening meal, and pack for our next destination—the bottom of the Grand Canyon.

☾

Hiking the Grand Canyon is like climbing a mountain in reverse. You hike in, all the way to the bottom, on the first day, from the high rim. The rim is where all the tourists go to look across the chasm that water and time carved through solid rock over millions of years, slow and steady. Up at the top, people pose at the edge to take pictures and stare. Up there it is safe. A precious few decide to walk down a hundred yards along one of the trails, either the Kaibab from the south rim or the Bright Angel trail from the north. The Kaibab is steeper but shorter than the Bright Angel, which takes its time snaking the shallow pitch angle along rock faces and formations.

Even fewer continue on past those hundred yards, out of view of the visitor center, the parking lot, the tour buses, until all they can see is the rock face to their right, the dusty—or, in winter, icy and slick—trail in front of them.

I've hiked the Grand Canyon twice so far in my life, both times when I was in college. Both times there were two competing selves that I hosted, as I placed one foot in front of the other over and over again. And those two people were as different as the two sides of Saturn's moon Iapetus, which hosts one of the brightest icy surfaces in the solar system alongside one of the darkest.

There was the part of me that thought, *Hell yeah, let's do this! I just finished three weeks studying star formation in irregular galaxies at Lowell Observatory! I am the youngest person here. I impressed everyone with my final talk! I can do anything!* That person's voice persisted even through the descent, which, though easier on the cardiovascular system, is hard on the knees and carries the greatest danger of falling.

After five hours, we'd made it down to the bottom of the canyon. We camped and side-hiked for a few days, venturing inside the shelter of Phantom Ranch in the evenings to sip hot cocoa and play cards. Then back outside at bedtime to roll out our sleeping bags beneath the stars. Down there, I felt strong, powerful, free. Far and away from problem sets and lapsed boyfriends. In pictures from our trip my hair is long, loose, and shiny, my skin clear.

On the day of our ascent out of the canyon, a different part of me emerged. This side of me I knew well, like a worn, soft jacket, or that old pair of Birkenstocks that had molded completely around my feet so that they were more comfortable than having no shoes on at all. This side of me said things like, *You're not getting out of this canyon on your own. They're gonna have to airlift you out of here or strap you to a mule like those lazy tourists. Only this mule will have to walk farther than any mule has ever had to walk in the canyon just to pick you up.*

Over the course of the next eight hours, as I plodded up through geological periods behind Joannah, sometimes in front, this voice got louder and louder. *Give up! GIVE UP! You're embarrassing yourself.* Where had my confidence gone? What had happened to that can-do attitude? It had scurried under a rock after the first switchback. I had no hope of finding it now.

I kept walking. The voice was loud now, roaring like a freight train. I kept my eyes on each foot coming around to step in front of the one I'd just placed down. One, then the other one. I kept my steps small and even. When I needed to stop and rest, I stopped, sat on a rock, and breathed slowly and deeply. I wasn't the only one. I know that now. Back then, even though others in our group stopped too, some as often as I did, I told myself that I was in the worst shape. Or rather, that voice told me, from that person inside me that tells me *I can't* and *I'm not*.

Joannah and I huffed and puffed our way out of the canyon, taking a couple of steps every ten minutes toward the end. The picture of us both emerging victorious at the trailhead, arm in arm, with our frame packs suddenly light as feathers, is one of my most cherished photos.

I made it out of the canyon. Once as a freshman, and then again, two years later, as a junior. Both times I thought that I wouldn't. Despite my training—hiking down and up Mars Hill every day of those three weeks at Lowell with my advisor, Jim—and despite being in my twenties and healthy as a horse, I swam in doubt. I didn't know then that those thoughts of defeat and shame came from parts of me that didn't have to define me. Those parts, sides, people, faces—they weren't all of me. I was something bigger, more expansive, that if I blinked I missed. On the way home from the canyon we stopped for the night in Monument Valley to have Navajo tacos and gaze at the rock formations made by wind and time. We woke up early to take pictures with the Sun rising behind the Mitten Buttes that had served as the backdrop of countless movies, especially Westerns. In the photos we took, the soft yellow and coral flames just breaking out of the Earth behind our shivering backs, there is no hint of doubt in my face. I look cold and happy.

The different faces of the Moon I see throughout each month don't change what the Moon actually is: a whole, complete, spherical body hurtling through space in slow orbit around the Earth. The Moon is bright in places, dark in other places, smooth in places, rough and cratered in other places. I see the Moon in light and in shadow. I know that without these parts of the Moon, the universe would be the less. I see the Moon whole, no part missing, no part detracting from its brilliance as it lies suspended in perfect union with the vast universe.

I am the Moon.

☾

I couldn't get enough of research. After Lowell, I headed back to MIT for my second semester and applied for a Space Grant to do a summer internship at the Jet Propulsion Laboratory (JPL) in Pasadena, California. I spent the summer working with Mike Shao in the Observational Systems Group "on lab," learning how to design a stable optical cavity that could be used in a high magnetic field. I mostly read optics manuals. Though the research didn't excite me, the environment did. I got to meet other scientists, many of whom were working on projects that would result in something flying into space. I met a NASA administrator who took me to a professional operating observatory—Table Mountain Observatory, in the Angeles National Forest, an hour-and-a-half drive from JPL. We watched comet Shoemaker-Levy 9 crash into Jupiter. Our summer student groups got to see a missile launch from Vandenberg Air Force Base. We toured Dryden Flight Research Center and the U.S. Geological Survey, where I got to meet Eugene Shoemaker himself—the man responsible for co-discovering comet SL9, along with David Levy. He'd also helped to train the Apollo astronauts to walk on the Moon and recognize rocks made of basalt—cooled lava—from those made of anorthosite, which was the holy grail, and what they were told to be on the lookout for.

When Apollo 15 astronauts James Irwin and David Scott picked up a calcium-rich rock covered in shiny crystals, Eugene's work was behind the discovery. The "Genesis Rock," as it became named, was originally thought to be a piece of the Moon's original crust from when it first formed. It turns out that the rock likely formed about three hundred million years later, after the crust had cooled and solidified. It had still been around a while, at just over four billion years old. Later anorthosite samples found during subsequent Apollo missions helped to truly date the Moon's age—about 4.46 billion years old, slightly younger than the Earth (about 4.54 billion years old). All of this helped to bolster the prevailing theory of the Moon's origin. The Moon was likely the result of some large thing—about the size of the planet Mars (which is half the size of the Earth and one-tenth of its mass)—crashing into the Earth shortly after its formation, ejecting material into orbit that gradually coalesced to form our Moon.

I saw every movie that came out in the summer of 1994, on the weekend that it came out. I lived in a two-bedroom apartment with three other girls working at JPL for the summer. We went to work, came home, went to the gym, and watched movies in big recliner chairs at fancy luxury movie theaters in Old Town Pasadena. We dined at my favorite restaurant in Old Town, Mi Piace. I ordered their Pollo al BBQ pizza every single time.

In the final weeks of the summer, my roommates already gone back home before starting the fall term at their schools, my stepfather Kirk arranged for me to stay with his friends' parents. The father was a retired JPL engineer who had worked on the Galileo mission. Galileo went to Jupiter's moons and sent back a ton of information, including indirect evidence of a salty ocean lying below the surface of Jupiter's icy moon, Europa. The mug he gave me, which has a sketch of the Galileo spacecraft on it, sat on the corner of my desk as I wrote this. It holds most of the pens I've used to draft what I wanted to tell you about moons, planets, stars, and galaxies.

It didn't matter where I did the research. It didn't even really matter what the research was. I was energized by it all. The following summer it was classifying hard X-ray sources back at MIT with Ron Remillard, while paying an ornery Irish male instructor from South Boston to teach me how to drive—something I'd never managed to make time for. The next summer, I conducted impact simulations into ice on Europa with Mike Nolan at Arecibo Observatory in Puerto Rico, using a model developed to test ballistic missiles at Los Alamos. I did this while pining after a surfer boy with long blond hair who was also a summer student there. When Hurricane Bertha came through, the summer students helped batten down the hatches at the observatory to protect it from the storm.

What mattered was how much I learned that I didn't know before, and someone taking the time to teach me that new thing. It's what I have to remember when I'm too busy to mentor an undergraduate because they don't know much yet and need lots of time and attention. Someone—many people—gave that to me once. My world, my very universe, kept getting bigger and bigger. Research became more a part of me with every project.

After Arecibo, my friends and I hopped around the U.S. Virgin Islands for the last few days before heading back to our separate universities. I almost missed the trip because I was scared. Instead, I gave in when my friends urged me on. I laid on the soft white beach on St. John and let the sand mold itself around my body. We strolled through the warm night air in the small village streets, alongside blond Rastafarian expatriates from the mainland. We stopped under tent restaurants to sit at tables with red-and-white checkered tablecloths and devour perfectly charred chicken dripping in barbecue sauce, letting the juices run down our fingers.

A few weeks after I returned to MIT, Jodie Foster landed at Arecibo to film the movie *Contact*. She drove up the gravelly dirt path around the circumference of what was the largest radio telescope dish in the world at the time, and yelled at Tom Skerritt's character for canceling her funding to listen for radio signals with the telescope. I would later watch the movie and recognize the new, large, golf ball–looking Gregorian mount suspended from wires and machinery above the center of the dish, and the thin scaffolding in front of it that I had stood on with a yellow hard hat to pose for a picture. It was all part of what had made that place special, evolving into an even more powerful instrument. Jodie Foster's character, Eleanor Arroway, had used Arecibo to look for life elsewhere in the universe. I didn't know it then, but that was only one way that it could be done.

Tide

The part of me that ached to learn everything I could about the universe was filling up. But another tank was dangerously hovering on empty. I could feel the imbalance like a lopsided car with too much air in the tires on one side and hardly any on the other. When I'd chosen to follow the path of astronomy, the part of me that I'd discovered at Exeter, that loved the creative arts, hadn't disappeared. I'd left it on the side of the road somewhere along the way. A year into MIT, it had caught up.

I needed more than science. I was starving for something—anything—that would allow the dormant parts of my personality to be a part of my everyday work. I wanted to use my feelings and opinions, rather than push them to the side in favor of facts and evidence. I took a course on international women writers at MIT. I was a Burchard Scholar, one of thirty-five students selected for a yearlong program of enrichment in humanities, arts, and social sciences. Leave it to me to find the one big liberal arts thing MIT did and go for it. Every month we were served a fancy multicourse dinner in a giant hall, followed by a lecture by a distinguished scholar. I ate it all up. I also took a world music class. I figured my parents would be proud. We listened to Tuvan throat singers. Jonna took the course with me. We were always so tired from late-night problem-sets sessions that it was hard to stay awake during the music listening part of class. Jonna was a master at listening with eyes closed, making almost imperceptible movements with her hands and face to convince the professor that she wasn't asleep, when she was. The class reminded me of a music class I'd taken in junior high. It was there that I'd learned the skill of being able to stomp and verbally count out whole notes, half notes, quarter and eighth notes with my feet and hands while simultaneously reciting sixteenth notes aloud. I can still do it.

I'd been singing in an all-female a cappella group at MIT since freshman year. People would file into the lecture hall and hum "Every Breath You Take" and "The Pi Song" (where we sang the nonterminating, nonrepeating number pi out to fifteen decimal places) right along with us. I enjoyed recording the first album the group put together. It was some kind of outlet. I thought it would be enough. But it didn't strike a harmonic chord in my bones the way acting had. I missed the story. I missed the silence between words, between lines, when a caught breath was full of some enormous meaning that I had to see the other character's face to understand. How someone could say so much without saying or doing very much at all. How there was a whole life in someone's eyes, if they did it right. Memories flashed through my mind, bringing back the giddiness I felt as I prepared to walk out onto a stage. Who was that person? Until I acknowledged her, anything else I did—science included—was simply another role to play, the mannerisms and habits of a character.

I took an introductory acting course at MIT. I guess I was cocky at first, because about halfway into the next semester, in a scene study course with the same instructor, she pulled me aside. She was a no-nonsense New York Jewish woman with straight, shoulder-length light-brown hair and large glasses. She commuted to Boston on a weekly basis to teach, and had a husband and young daughter at home. She'd been talking to us about the play *Angels in America*, written in response to the devastating HIV/AIDS epidemic. It had won the Tony Award and the Pulitzer Prize. She told us about how one of the characters had said a line to another character, a man who was sick with the illness. She'd told him in the gentlest way that she could see a part of him that was untouched by disease. The way this actress had said the word *part*—touching it ever so lightly, as if the word was a piano key depressed by her pinky finger—meant all the difference in the line. The moment between the two actors was pure, simple, and tender. It disarmed everyone in the audience, breaking them open like a piece of ripe fruit. All they could do was cry.

We all went through part of our monologues. Mine was from the movie *When Harry Met Sally*. It was the one where Meg Ryan's character, Sally, tells Billy Crystal's character why she and Joe broke up. How neither of them wanted kids at first, but then one day she was playing "I Spy" with her friend's daughter, and the little girl said that she spied a family. Sally had started crying.

After class the instructor came up to me and told me that I had done well. I wish she'd left it there. But she didn't. She said that I appeared to have shed the "air of pretension" that I'd had, and was now being real and authentic.

I smiled and thanked her, then left the classroom and headed back to my dorm. My feet grew heavier with each step. I felt like I did when I was eight years old and a boy at school out of nowhere came over and punched me in the gut. In the space between the blow and the pain, my breath caught a gust of wind and sailed right out of my body.

Once my breath returned, it spread throughout every cell, igniting a fire inside. I fumed. I had no idea that I'd been behaving pretentiously. To be pretentious seemed like a conscious choice—a choice to present to the world a face that said, *I am hot shit*, whether that was the case or not. Maybe subconsciously I thought I was hot shit because I had won the drama prize in high school. *Give me my Oscar, thank you. I'm ready for my close-up.* But I didn't strut around. I didn't heckle others when they made mistakes. I wasn't a jerk.

I decided that day that I hated that word—*pretension*—and always would. Greater than my hatred, though, was my fear of the word, and the label, even retroactively ("you used to be this way; now you aren't"). I never wanted to seem arrogant, cocky, "too big for my britches." I wanted to find a way to make sure no one would ever have cause to use that term to describe me again. If I was small, unsure, returned compliments faster than I could take them in, that might do it. Pretentious people basked in their own glory.

Senior year rolled around, and it was time to decide what was next. Shelly had been accepted to Harvard Medical School and would be staying in Boston with her boyfriend, John. Jonna was staying in Boston too, to get her combined masters in engineering and business at MIT. I applied to five or six astronomy graduate schools and three acting grad schools—Yale, NYU, and the Globe at UC San Diego. I shot for the Moon all around. Maybe I was pretentious.

I didn't get into any acting schools. But I was offered a full scholarship to the PhD program in astrophysics at the University of Wisconsin–Madison.

Work I did with Deidre turned into my senior thesis and a published paper with galactic astronomer Bruce Elmegreen. The paper, "The Relationship between Gas, Stars, and Star Formation in Irregular Galaxies: A Test of Simple Models," is still one of my most cited papers. In it, we determined that the ratio of observed gas density to the critical density needed to form stars is a factor of two lower in irregular galaxies than in spiral galaxies. This was the reason why it was harder for irregular galaxies to give birth to new stars, compared to spirals. It was not a theory anymore.

I arrived in Madison in the late summer of 1997. I remember walking into a dark room in the department where three or four students sat in front of computers. They were remotely observing, telling a telescope somewhere else in the world which galaxy to point to next. On a television in the corner was coverage of the car crash that killed Princess Diana and Dodi Fayed.

There was a blonde woman—a third- or fourth-year grad student—who seemed so confident and sure of herself. She'd finished classes and was the one doing the observing. UW–Madison was well-known for galactic astronomy. John "Jay" Gallagher, the department chair, was an expert. So was Linda Sparke, a small but mighty spitfire of a woman with short, spiky hair dyed red, and a strong British accent that she used largely in service to irreverence. She was brilliant, so this was okay. No one could argue or fault eccentricity when you had the intellectual chops to back it up.

I'd received a full fellowship, with no teaching assistant requirements, so that I could focus on my courses. They knew I'd struggled with physics courses at MIT, and that I'd also done a lot of research. They wanted to help more Black people succeed as astronomers, before it even became fashionable to want that. They even had one Black faculty member—a rising star named Eric Wilcots who specialized in mapping neutral hydrogen in spiral galaxies. Neutral hydrogen was everywhere in the universe, had been there since the beginning, when things started to cool down after the Big Bang. Mapping its concentrations could tell us something about how galaxies moved through space and prepared themselves to make new stars. I would work with him, continuing my study of galaxies, but this time focusing on spirals—galaxies that looked like our very own.

Atoms of only one proton bound to one electron, one after the other, throughout expanding space. That's all that's in a hydrogen atom in its neutral state. One proton and one electron. The proton hangs out at the center of the atom alone, without a single neutron to keep it company. The electron orbits in a shell around it. The electron isn't particularly full of energy it needs to release, or too excited by hot stars or collisions with other particles to stay in its orbit around the proton. Everything about a neutral hydrogen atom is inactive and unremarkable. Neutral.

This is how I felt about being in Madison. Neutral.

I wish I remembered more about that year. My memory is like a photo album, full of images out of order. The images prove that I was there. But everything else about my experience points to someone else's life, someone who was tired and weary after four years at MIT and needed a break. But I didn't know how to listen to myself yet. I don't even remember doing any actual research during that year.

The electron in a neutral hydrogen atom, barring a few possible outside events, remains anchored in its commitment to revolving around that proton. An electromagnetic force keeps it attracted to its oppositely charged central proton, keeps the two in orbit around a common center of mass between them.

Without a common center of mass, a moon will not revolve around its planet, nor will that planet orbit around its host star. Rogue planets do exist, without stars to warm them. But these planets hurtle through space in darkness as cold and barren wastelands. I'll tell you more about them later. Planets, like ships and people, need a force of attraction to anchor them. The force between the Earth and the Moon keeps both together, bringing us tides to rise and swell up in the oceans. The Earth never gets carried away by them, though. The tides are strong, and the Earth is too. The force between the Moon and the Earth creates that powerful anchor. Anchors hold bodies firmly planted in time and space.

Anchors connect a body to the ground beneath, to something bigger that guides and protects, filling it from the inside out. Anchors prevent bodies from hurtling coldly through space into black holes of loneliness from which there is no escape.

The three other first-years in my cohort had written to me in advance to ask if I wanted to live with them in a four-bedroom house they'd found not far from school. I thanked them for the invitation, and declined. I wanted the freedom to play the music I wanted to, watch whatever TV shows I wanted to, walk around the house naked if I wanted to, and call friends and complain about all of these people loudly and often. There were far too many reasons not to live with my classmates.

I rented the top floor of a house under renovation, sight unseen. I showed up in a cab, and the landlord walked me through the two bedrooms, bath, kitchen, and living room with lots of bright sunlight pouring in from the front windows. He'd kept the apartment furnished at my request. He gave me a set of keys and left me to unpack my two suitcases. It was way more space than I needed, but I was fine with that. It was all mine.

As I went to put sheets on a twin bed in the bedroom I'd chosen to sleep in, I noticed something scurry between the two slats of the bed frame. It was black, shiny, and about half the size of a dollar bill. I froze.

I didn't sleep that night. Instead, I spent the night hunting the two large beetles that had not signed on with the landlord to rent out this floor. They were the original occupants, and as far as they were concerned, the house was not for rent.

The rest of the year was like that. The landlord tried to help—with bug bombs, sprays, and d-CON to kill the rats that had also made an appearance through the walls. The d-CON must have worked, because a foul smell made the living room light less desirable to bask in on the weekends. Almost every night I would walk into the bathroom, turn on the light, and see a new insect lounging in the basin of the mint-green sink.

I can't fully explain to you why I didn't move. I don't even remember looking for other apartments. The only way I can describe it is that I was hanging on to something that I wanted to work. I wanted to make the choice not to live at "Astrohaus"—as my classmates living in the house referred to it (one of them was German)—worth it and right.

I befriended a few people who lived in a co-op, full of eclectic people from all around the world, some of whom were studying or working at UW–Madison. Their rooms were full of thick rugs in paisley patterns. Different languages and accents filled my ears as I roamed around the halls. I spent every free moment there. I joined their quilting bee. I joined the co-op's cooking rotation and made vegetarian stew for the house members. It gave me something to hold on to that had nothing to do with astrophysics or roaches. Of course I developed a crush on a white man who lived there, and told him so. Days at the co-op became awkward after that.

I paid $600 a month for my two-bedroom top floor of a house. My fellowship stipend was $1,000 a month. Putting 60 percent of my take-home pay toward rent seemed like a justifiable percentage for my freedom. It left me $400 for the entire month for groceries, clothes, transportation, movies, and eating out with friends. I could rationalize my way out of some of those things. It was too cold to go out for fun. Who needed to put money away in savings for the future? I was twenty-two. I had plenty of time for that.

I started putting groceries on credit cards. I did need to eat. I lied about my annual salary on credit card application forms to get a higher spending limit. When the credit card companies started calling, I told them that I was not home, but that I'd be happy to take a message and have me call them back.

I started to daydream. I would sit down to work on a problem set, and a great line from a movie would pop into my head. Scenes I remembered watching on-screen in dark theaters would replay themselves in full color. The moments in those films would stand out like anthems sung in full-to-bursting stadiums—timeless, suspended in the crisp air of a coming fall, when the world holds its breath with the hope of a fresh new year. In those perfect scenes of words, eye contact, music, and lighting, everything good seemed possible. That is where I stayed in my mind.

I was on the red carpet. I'd been nominated for an Academy Award for Best Actress for playing the title role in the highly acclaimed Zora Neale Hurston biopic. I won the Oscar and gave a heart-stopping, tear-inducing acceptance speech while dressed in a gown of bronze crystal etching over thin, transparent muslin. It was a dress that stood up to even the harshest scrutiny on the *E!* network. My thoughts full of stories, I found an acting class on campus and started attending.

Things weren't always hard. I aced atomic physics. It was really a series of rules about the orbital shells of atoms and how particles moved within them that, when memorized, made things pretty straightforward. But mostly, I couldn't focus. My grades in Stellar Astrophysics and Basic Astrophysics suffered. I had nothing to anchor me beyond school. So when school started going badly, I was adrift. In my upstairs two-bedroom half of a house on a deserted Madison street, I felt cold and lonely, neck-deep in snow inside, to match the howling winter on the other side of my windows. A professor told me to consider other career options. He was white and old, with a few thin wisps of hair erupting from the top of his head, and bulging eyes that surveyed me through half-inch-thick lenses. I sat in his office and cried.

With no internal anchors to keep me grounded amid my academic turmoil in the PhD program, the excited states I inhabited came from outside forces. A movie someone mentioned seeing. A new restaurant refuge

I stumbled upon on State Street. A co-op crush. Nothing lit me from within.

The mind is a powerful thing. It could trick me into believing anything. It just needed to find evidence. No astronomy graduate students with brown skin within five miles of my department? That means there aren't any, and shouldn't be. No fashionably dressed female scientists at the head of my classrooms teaching, or in the media? They're like penguins and polar bears, fashion and science. Never the two shall meet. My own mind became the biggest racist I had ever known. I told myself that the professor was right. I wasn't cut out for science, for a career in astronomy. Nobody who was a scientist looked like me. That must mean something. My research advisor was Black, but he was also a man, so maybe that's why it had worked for him. I was a Black woman. I liked makeup. There were no astronomers who cared about their looks.

Astronomers cared about drinking though. The grad students took over an old observatory that the department operated in the country and had a party. The faculty attended for the first hour to show support. I stood at the kitchen table with my classmates and professors, a shot glass full of tequila in my right hand, a sprinkle of salt between the thumb and lower knuckles of my left index finger, and a lime on a plate in front of me. I was sure that our professors would say to us at any moment, "How dare you drink in front of us!" But they took their shots too, and after some small talk, scurried out with family excuses and left us alone.

I woke up sitting on a toilet the next day. Red wine and tequila do not go together.

Something about the loss of control—letting go of the worry over my grades, and the constant question of whether I had made the right choice to come to graduate school, which I'd also stopped entertaining as a result of getting absolutely shit-faced—had been a relief. The only problem was that Gloria and Ken and Nancy and Nick and Zack had danced and partied and gotten drunk to blow off steam as astronomy graduate students. I'd done it because I knew I didn't belong there and I wanted to forget that. If I wasn't forgetting it, then I was going out of my way to point out just how different I was. I knew something about this or that movie. *Oh yeah? You got the answer to that homework problem in stellar astrophysics? Well, I am taking a scene study class in the drama department, and I love it.* I did. It felt good to be good at something again, and to be told by my acting instructor and classmates that I was good. No matter how much my astronomy professors knew about their fields, they didn't know about acting. Not enough to tell me I couldn't cut it there. If they'd wandered across campus into my world of acting, I had an edge over them. In that world, I was a king.

Migration

Once I faced the hard fact that I looked forward to my acting class at UW–Madison more than any of my astronomy classes, I started to plot my escape. I applied to acting graduate schools again. I secretly rode buses to Chicago to audition. I auditioned for the National Theatre Conservatory in Denver, Colorado. It was a standalone MFA program, not affiliated with a larger university. The students worked as actors in the conservatory productions as part of their training, so that by the end of their graduate program they left with their Actors' Equity cards, which meant union membership. And, on top of all that, the MFA program was fully funded. There would be no student loans. I had enough student loan debt from MIT, even some from Exeter. I didn't need any more.

I did a monologue from *Henry IV, Part 2* for my audition for Denver. It was the scene where Lady Percy begs her father not to go to war, using every persuasive maneuver in her arsenal, including guilt-tripping him for leaving her husband without backup in a previous war—in essence blaming him for her husband's death. It is a beautiful monologue, one that every mature woman dreams of getting when doing Shakespeare. It gave me the chance to show anger, grief, vulnerability, deep sadness, and overwhelming pride.

After I finished, the man running the audition told me that I reminded him of "a young Miss Bassett." He meant Angela Bassett, whom he'd apparently known from her days at Yale. "But you need training," he said. He told me he wanted me to come to Denver to visit. In other words, I was getting a callback. But for this callback I would get on a plane.

At the time I wondered what he meant by "training." What had he noticed? I thought I'd spoken all right—well, even. As much as I was there in hopes of being offered a chance to study acting seriously, I had no idea what that actually meant. I thought that someone might tell me how to be more convincing as a character, by using a certain look or mannerism. Really, I hoped they would see how amazing I already was and give me more parts to play, leading to some production exec coming to one of my performances and casting me in their next movie on the spot. What I didn't know, had given absolutely no thought to until that moment, was that acting—like any other art form, like music or painting—required a sharply tuned and refined instrument to do well. Almost no one could stumble in off the street with no training and be instantly spectacular. More often, soaring as an actor required wings made strong by years of practice, day in and day out. This was like anything else in life. It was true with astronomy and physics too. Why had I not thought it to be true about acting? Maybe it was because when it was done well, it looked effortless.

I'm thinking of Sally Field in *Steel Magnolias*, when she's at the cemetery after her daughter's funeral, and she loses it. But she doesn't try to lose it. She tries to keep it together, after losing the one thing that she could never lose and still want to live, the thing that every mother fears above all else actually happening. She tries to keep it all in, but all of the well-wishes and attempts at verbal comfort and prayers just send her over the edge, and the world explodes out of her.

I got no sleep the night before my flight to Denver. I had a problem set due in one class and an exam that day in another. I showed up as a hollow shell to my callback audition with a group of other prospective students gunning for a spot in the MFA acting program of our dreams. I went through the motions. But I wasn't myself, and I knew it. I blame the homework and the exam, but who knows. The spaces they showed us were beautiful, especially the theater in the round. The stage was in the center, completely surrounded by seats, so that the audience could see the action from all angles over 360 degrees.

I received my rejection letter a few weeks later. I also received a second letter, from an older instructor I remember meeting at the callback. It read, "You have a rare talent. Nurse it carefully. You have a wonderful career ahead." It was the exact amount of sugar on the teaspoon that I needed to help it all go down.

The next weekend I took a late bus down to Chicago again, this time for an audition for UCLA's MFA acting program. When I got there I was told that I would be in the last group. I wore a flowing, sleeveless hippie dress in dark paisley patterns that I felt would work with either my classical Lady Percy Shakespeare monologue or my contemporary one from *When Harry Met Sally*. A man opened the door of the audition room on the hotel's main floor, and a group of twenty-somethings filed out. He stood there in baggy striped balloon pants, with a dark T-shirt and a short-sleeved button-down shirt hanging loosely over it. He said his name was Tom. He invited me inside, along with a guy who was also waiting. We were the group—the last one of the day.

We got inside and Tom put two chairs facing each other in the middle of the room. He told us that he was tired, and that rather than put us through the whole spiel he had been doing with the other groups—with movement exercises and everything else—he wanted us to sit opposite each other and do our monologues. Not perform them; just say them to each other, simply, with no effort. Let them fall out of us. I did the Meg Ryan monologue. It seemed the most conversational, the most intimate. In the movie she and Billy Crystal are sitting opposite each other at a small dining table in a dimly lit restaurant, talking.

As I got ready to leave the hotel lobby for the bus that would take me back to Madison, and back to regular life as a lapsed astronomy grad student, Tom came over to me. He asked if I would ride up in the elevator with him. It occurred to me sometime after the elevator doors closed that this could go horribly wrong. But something inside me trusted that it wouldn't. Still, I made the decision not to leave the elevator with him if he asked me to.

Tom started off with disclaimers. The UCLA acting program had no money and hardly any space. Students rehearsed in the parking lot, the sculpture garden—wherever they could find room. The hours were long—classes all day, rehearsals in the evening. Then he said, "I want you to come to UCLA."

He told me I would be getting a packet in the mail. He got off at his floor, told me he hoped I would decide to come to UCLA, and left. I stood there as the elevator doors closed again, as the elevator sank, falling floor after floor, feeling queasy, light-headed, and skeptical. How many students had he ridden up in the elevator with that day? How many had stood there where I stood, watching him walk down the hall in his balloon pants, wondering if he was right and they really were special?

My acceptance packet came a few weeks later. By that time I'd already decided that I would go to UCLA if I did get in. I didn't care how many loans I would have to take out. This was my ticket out. I wouldn't miss it.

☾

The summer after leaving something behind is a special kind of fortune. I carried no weight, only the dim memory of a place in the middle of the country that I visited as a tourist. The memory of the place began receding even before I left.

I requested a yearlong deferment from UW–Madison. It would give me a year to determine whether the acting thing was for me, or if astronomy was really what I wanted after all. That's what I told myself. But deep down, I knew I wouldn't be back. I gave my landlord notice. I told my classmate Nancy, who'd been a good friend. She'd driven me to get groceries in the winter, and we'd even taken a trip to see her family in the Upper Peninsula in nearby Michigan. She was focused, clear on being there to get her PhD in astronomy as training for whatever came next. She offered to ship boxes home for me. I would stay in intermittent touch with her for a while, even write letters. I would one day apologize for the way I was that year, some of it harmful to her, like gossiping about her to try to get yet another white guy to like me back. I don't remember why I would have thought that was a good choice. I was incapable of judging the good ones from the bad ones back then. The thing everyone says about misery is true. I was so used to standing in the way of my own light that I couldn't stand the light coming from anyone or anywhere else.

On the bus leaving Madison headed for Milwaukee International Airport and back to Groton, Connecticut, to my mom, stepfather, grandmother, brother, and sister for the summer, I was free of the crushing weight of my twenty-years-long aspiration. It was an odd sensation, this lightness, as if I might rise up into the air at any moment. I knew leaving Madison meant leaving astronomy, probably forever. The only part that stung was that I hadn't left it on a high, as my brief English MIT boyfriend had always suggested. When I'd started to struggle with courses at Madison, I'd reached out to Deidre for support. She'd written back telling me that courses were what you had to get through to get to the fun part, which was the research. I hadn't hung in there for that. I'd gotten off at the nearest rest stop. That dull ache would sit in my stomach and wait for the right time to bloom. Right now, physically separated from all that reminded me of failure, I felt like a whale set free in the ocean after so much aquarium life.

When planets form, they don't always stay put. Other planets come close—sometimes too close. And their mutual attraction can break them apart, or fling one of them out of its comfortable, familiar orbit and into oblivion. Planets can migrate toward their stars, far closer than their distant natal origins. This is one possible pathway for a special class of planets called "Hot Jupiters." The gas in their atmospheres had to have formed in cold conditions, farther out from their stars. Yet now they sit in suicidal orbits a relative stone's throw away, willing participants in the excision-by-fire of their atmospheres. Some say that habitable cores of these planets could remain after their entire gaseous atmospheres have burned away. Maybe beneath will lie what was always supposed to be there, something that would never have formed had these planets not dared to leave their places of birth in search of something different from what they had to begin with.

At the end of September I showed up in Los Angeles with two suitcases and a credit card. I stayed at the UCLA Guest House on campus. I gave myself five days to find a place to live. I bought a good pair of sneakers and walked each day to the community housing office at the other end of campus to look for an apartment. On the fourth day I found a room in a family's home less than a mile from campus on a quiet street in an expensive, tree-lined neighborhood in Westwood. I bought a bicycle. I went to the movies down the street at the Century City Mall, $4.50 for a ticket, killing time until school started.

My room had lots of sunlight, plush light-green carpet, a twin bed, large desk, dresser, mini fridge, and microwave. The family requested that I not use their kitchen. The boxes of pots and pans Nancy shipped to me from my Madison roach motel stayed in the closet. There was a small TV on the dresser that I used to watch episodes of *Friends* until school got busy. The bathroom was right outside my bedroom door, and I had private access to it. There was a teenage daughter in the house who I never saw. I was the perfect tenant. I always paid on time, with money from the student loans I'd taken out to be there. And once school started, I was never home. The only sound that could be heard by others in the house was the loose clicking of the spokes turning over inside the wheels of my bicycle as I guided it through the entryway, across the living room hardwood floor, past the stairs leading up to rooms I was not privy to, and into the carpeted bedroom that was all mine.

I'd accepted every bit of financial aid they'd offered me. To me it was free money. I knew I'd be in school full-time, so there was no point in trying to get a job to offset the loans. I needed to be rested and at my best for my acting program. I didn't have a computer to write papers on, so after a few weeks of using the computers at Powell Library on campus, I took out a separate computer loan from the school financial aid office and bought myself a desktop PC. These were all justifiable expenses given my present circumstances. And small prices to pay for a shot at the new dream I had nurtured and kept warm in the snowy winter depths of Madison.

Acting

Life on Earth requires three things: liquid water, a suitable environment to make organic molecules from elements essential to biology—sulfur, phosphorous, oxygen, nitrogen, carbon, hydrogen—and an energy source.

Liquid water is the most elusive of the three requirements for life as we know it. A terrestrial planet by nature has some kind of energy source—whether it's stellar or chemical—along with the basic building blocks, in some form, that are needed for life. What isn't as common—at least in our solar system—is liquid water. That's because its presence depends on a very particular combination of temperatures and pressures that are required to keep water in liquid form. Life on Earth uses a diversity of metabolisms. And they all require liquid water—for chemical bonding, and to dissolve other substances for chemical reactions. So, in keeping with the slogan of astrobiologists—"Follow the Water"—it is liquid water that drives our search for life elsewhere.

Artistic expression was my essential element. I hadn't realized it until I was fully immersed in it, day in and day out. In graduate school for acting, I came alive. It was playtime compared to grad school in astrophysics. That isn't to say that it wasn't hard. But it was hard in a different way—one that didn't make me think as much. That was the difference. In the world of the theater, thinking was the obstacle. I was supposed to feel, and to let myself be guided by my feelings. Shakespeare was taught and described to us as "living thought." There was no subtext, no subconscious, thinking one thing and saying another. The characters needed to appear as if they were talking from their first thoughts—nothing censored, nothing held back. So the actors needed to too. My scientific brain, the one that analyzed and considered and would scarcely admit being unknowledgeable about something, needed to shut itself off. In this business, unlike in science, wearing my heart on my sleeve was a good thing.

My instructors taught me to use my whole body to communicate and to let my voice resonate throughout my body's cavities, from the deep lower abdomen, through the intercostal muscles between my ribs, all the way up through my nasal cavity, and even through my scalp. I had been using the tiniest fraction of my body—from the neck up—to communicate. There were so many places within me where power lay dormant, waiting.

The personal aspect of my acting classes was revolutionary. It nurtured me in a way that I had been missing so deeply in my science life. In one class, Judy Moreland instructed us to choose a poem and perform it, with all of the voice and speech training we had acquired from her so far. In an interdisciplinary class called Collaboration, taught by the chair of the department, Mel Shapiro, our job was to bring in pieces created from scratch to show him. Four of us created an entire series of scenes where the only word any of us was allowed to say was *blah*. Serious, funny, urgent, romantic—we conveyed it all, uttering only that one word. It was a powerful exercise in using all that existed in the world beyond words, and understanding how much of an impact we could still make without them if we worked hard enough.

There was another piece that I presented to Mel on my own. I wore a long, chocolate velvet ballgown and black heels, and stepped in front of the class with my violin and bow. I held my instrument underneath my chin as if poised to play a marvelous concerto. Once my bow touched the strings, what emerged were the screeching, sawing sounds of a toddler during her first lesson. At one point during the "piece," I plucked a single string as if it had been written into the imaginary stanza. The class roared with laughter. Mel laughed so much that he cried.

In our method acting class, we had to do these things called "private moments," where we performed two activities we would do in our private lives that we wouldn't mind others seeing us do, and one that we would be mortified if someone saw us doing. I remember that I flossed my teeth in front of the class, staring into a make-believe mirror. My classmates told me later that they thought I must have been a dentist before training as an actor. But I had grown up answering the same question on the phone each week from a father who'd had dental issues in his twenties: "Are you taking care of your teeth?" I had finally started flossing to get him off my back.

For my embarrassing "private moment" activity, I sang my lungs out to "One Song Glory" from my favorite musical, *Rent*. I remember that my instructor, Salome Jens, told me after I'd finished that I'd almost gotten there, almost gone 100 percent into my own world and sung like no one was watching, even with the eyes of the other eleven members of my class on me. This was the whole idea behind the private moment exercise—to get us to learn how to let our walls down, be vulnerable, let others peek into our characters' lives, raw and unfiltered, so that any audience would see into our soul and become changed by that intimacy.

I think I did pretty well for a recovering scientist.

I would be lying if I said I don't replay that private moment over and over in my mind when I'm alone and dancing. I imagine all of the members of my class sitting in a row of chairs in front of me. They are smiling, their toes bobbing up and down to the music because they can't help it. They watch me move, unaware of them. I don't seem to care if they are there or not. That's why their necks crane to be a part of my world. They are mesmerized. I've nailed it, and they know it.

I lived in sweatpants and a T-shirt. I could move in those clothes, and that's what I was supposed to do. My instructors didn't just care about my head. They wanted all of me, including my shoulders, which I had to learn how to hold like Natasha from Chekhov's *Three Sisters*, or Blanche from *A Streetcar Named Desire*. They wouldn't hold their bodies the same way, because they were not the same person. One was proud, the other scared. Perhaps they were both scared for different reasons. Finding out what those reasons were, and infusing that backstory, that "moment before" into everything the character did and said onstage was my job. There were no problem sets. But there was homework.

I learned what that man from the Denver conservatory had meant when he said that I needed training. Here, in acting school, my mouth learned how to form words and phrases the right way. It was a way that would conceal from the audience where in the world I, the actor, had grown up. We became blank canvases to build a character on. On top of those canvases we could put a Southern or a British or an Australian or a New Zealand accent. Or we kept the canvas "blank," so that we sounded like what the acting world considered neutral: upper middle class, educated, and well-read. Or we intentionally changed a diphthong or a consonant sound here and there to sound more colloquial, informal, less wealthy, conveying a "General American" dialect rather than the elite "Standard American" dialect. I knew how these differed, because I was taught both. I memorized the proper shapes my lips formed to make each vowel sound appropriate for the corresponding dialect. My tongue curled and shifted itself up toward my front teeth or toward the back of my throat as needed. I was a good student. I had a good ear, from my violin conservatory training. I got all the sounds right on the test.

My body swayed and curled and unfurled to the music in movement class. Tom was there in his balloon pants, and he carried a large stick like Debbie Allen did in *Fame*, to signal a change in tempo. He taught us all what we could do to stand out from a crowd by just being ourselves rather than following a leader. This was good training for me, ever the good girl. One time in his class I played the drums while a female friend and classmate danced in front of me. I was focused on her, she on me. We were embodying the music, ignoring everyone else in the room. Tom told the other students to watch us. I thought he was spotlighting my friend. She was the one dancing. But then he said, "Who are you all watching, class? The dancer or the drummer? The drummer, right?"

My friend sobbed in the bathroom afterward. I felt awkward, uncomfortable, and oddly proud at the same time. It was the risk we all took in acting, in being here. At times we would be the example of the right thing to do, and we would shine for it. At other times it would be the opposite, and my girlfriends and I would drown our sorrows in chicken Madeira and avocado rolls at the Cheesecake Factory in nearby Brentwood, complaining about our eccentric instructors.

I would have my sob-story moment too, when I showed up for my yearly review by the faculty. I wore a blazer to the review. It was a formal occasion. It wasn't the time for sweatpants. Tom saw it differently. One look at me and he said, in front of the rest of the faculty, "If you'd worn a blazer to your audition in Chicago, I would never have admitted you." I crumpled, slinked out of the room, and wailed.

Then I got mad. I stormed back into the room and told Tom and the rest of the instructors, cry-yelling, that they were lucky to have me in their program. Then I left again and cried some more. The chair got wind of what had happened and called Tom. He brought me a pie the next day.

My energy was boundless. I was in my twenties, set free to explore my entire body and breath, use their power to become character after character, rest, then get up and do it again the next day. I told many stories, became many characters as an actor, onstage and onscreen. All of the personal work, the mining of my own experiences and bringing up my own history so that every emotion I had ever felt was fresh under the surface, like the young features of Jupiter's ice moon Europa, helped me learn who I was.

Our "coming out party" was a showcase to which agents were invited. It was nerve-racking. Some of us got an agent out of it, and some didn't. The biggest moment of the program was something else entirely. In our final year we wrote and performed one-person shows for our master's thesis projects. The coursework our instructor Tim Miller had us do involved writing the stories of our lives on long scrolls of paper and covering our bodies with them. We literally wore our personal histories for all to see. It was a way of getting comfortable with being exposed and vulnerable, and discovering the power in that vulnerability. No turning away from our stories—our childhoods, our innocence, our corruption, the disappointments and the victories. We learned to accept and own it all.

My solo show was called *Goddess. Divided.* It was the theatrical embodiment of my brain and all that went on in there. In one scene, I wrote equations on a chalkboard while giving an astronomy lecture. Throughout the lecture I exploded into famous lines from movies—one spoken by Marilyn Monroe in *Bus Stop*, another by Kiefer Sutherland in *The Lost Boys*—as if I had Tourette's. I got the point across.

Home

I remember when I met Steven. He was wearing yellow-tinted sunglasses with black frames. His hair was sandy brown and scattered around his head. He hadn't shaved in a few days. He was sitting. I was standing. He looked up at me through his horn-rimmed shades and smiled a big smile. Later he would tell me that he thought he looked like the actor Vincent D'Onofrio. Sometimes Jack Black. His mom would tell him he looked like the baseball player Mike Piazza. I saw all of these men in his face at times. But his smile—the one that's full of mischief and proud as a peacock—is all Dennis Quaid.

I watched him kick around a hacky sack in a tie-dyed shirt and cut-offs. His calves were defined and golden brown. His shoulders were broad. There was a lightness about him. It seemed like he was aware of the world and accepted it, and his place in it. There was no apology in any part of his body. He seemed to know who he was. And every so often his eyes would dart to me and then away.

That first afternoon, after orientation, we all went over to our classmate Mike's place to watch baseball. He and I were given the job of walking to the grocery store for snacks. We had our first argument on the way. It was about the movie *Saving Private Ryan*. He thought it was sensational. I thought, with the exception of the first fifteen minutes on the beach at Normandy, the movie was just okay. We stood in the checkout line like an old couple even then, familiar. Standing or walking together, in class, at lunch, there was a comfort I had only known with my oldest friends. Words fell out of our mouths with ease. We were comfortable in silence. He was becoming someone who had always been there. I didn't have a word for what this was yet. I knew that it was friendship. But it was also something else, something I had never had before.

One morning before method acting class, as we stood outside the classroom waiting for the door to open and students to pour out, Steven told me my shoe was untied. I looked down at the loose laces and stuck my foot out on the ground toward him like a queen. I had no idea what he would do. I didn't expect what he did, which was kneel down silently and proceed to tie my shoelaces. His humility stopped me dead. I stared at him as if he were a famous abstract painting at the Louvre. *Who was this man?* . . .

During one of our three-hour evening phone calls, before we fell asleep on the phone with each other, I told Steven my greatest fear—that I would be murdered in raggedy underwear. He didn't miss a beat, gave me his best impression of Inspector Clouseau surveying my dead body in the street with skirt lifted, bemoaning the state of my panties. *"Pauvre femme . . ."* It was that night that I fell in love with him.

I wasn't sure about it all yet. At a club he belted out "I wanna fuck you like an animal!" along with Nine Inch Nails. He was sweaty and hairy and I could tell that he knew about sex. But in his car in front of my house we sat and listened to the Goo Goo Dolls, and he sang gently and sweetly, and I sang with him and I could have stayed there forever. When a few of us went out to a bar late one night and he wasn't there, I only wanted him to walk up the stairs. I pictured it, tried to will it to happen. He would be wearing a crisp white shirt, black jacket and slacks, and his sunglasses. He was too cool for it all, but he was there for me. He never showed up, and that night I chose him without him knowing it.

On the beach on our first date, I told Steven that I was a virgin. I told him I'd had opportunities to change that status but had chosen to wait until I could see my child's eyes in his. He thought that was beautiful. He probably also thought that I was a prude, at twenty-three years old. But he never told me so. In the movie *A Night at the Roxbury* we held hands, tracing their shapes and lines with our fingers. I have no memory of the movie. But I still have the ticket stub. (We were married exactly six years later, to the day.) He'd opened the car door for me like he always did, in front of the home where I lived, and walked me to the front door. He'd asked me if he could kiss me on that doorstep and I'd said yes, and he'd kissed me and I'd let him, and then my knees got shaky and he held me and I stayed there and didn't want to leave that doorstep. We were inseparable after that.

During that first quarter, the only quarter in the entire two-year program when there was no show to rehearse at night, Steven, Robert, Sarah, and I joined the campus ballroom dance club. We were supposed to switch partners after every song. We never did. We were that couple that kissed on the stairs, on the benches, in the movies, in front of all passersby. No one else existed. Sometimes someone would tell us to get a room as they passed. We didn't care. Steven held my hand, kissed me, all of it, in public. He was white, from Nebraska. I was Black, from everywhere—Southern California, Canada, Massachusetts, West Africa in my blood, and Ireland too. He was never self-conscious. Sometimes I was. But his certainty, his pride, made me brave and proud too.

On Fridays after rehearsals we rented our favorite movies of all time to show each other. We had been roaming the world for twenty-plus years before finding each other. We had some catching up to do. Sometimes we rented five movies neither of us had seen from a neighborhood video store. We stayed up watching all of them, imagining ourselves as the characters, in Steven's studio apartment a block from Mann's Chinese Theatre and the Walk of Fame. It was a seedy section of Hollywood, but I was safe with him. We'd stay up until 5 a.m. watching movies and eating stew with extra dumplings he made from Bisquick, or order from Nat's Thai because they delivered until 3 a.m. We ate German chocolate cake and cinnamon cream-filled pancake triangles he made with a sandwich press. I called them "crumb fancies." In the early morning hours we fell asleep and woke up after sunset, made dinner, and did it all over again on Saturday night. We had a word for what this was now. We both used it to describe the other. We'd say, "You are my everything." We knew we were separate and whole on our own. We had each done our own work to know that. It made us even stronger together, instead of weaker.

I would come out of acting class and look for him. I'd find him sitting on a bench outside the theater department buildings, smoking. I'd never known anyone who smoked. Steven made it look so sexy. There he'd be, puffing away, and then his eyes would find me and that wide smile would spread across his face, and inside I would jump up and down. He always made the smoke go somewhere other than in my direction when we were in a crowd. When it was just us, he put the cigarette out. Later, when his father got sick, there were Post-it notes on his bathroom mirror and his fridge and his computer monitor screen. On all of them was the same single word: *Emphysema*. I was thankful when he stopped smoking. I also missed the bad boy on the bench puffing away without a care in the world. I knew that was selfish and vowed to never tell him.

Temping

I'd scored big getting signed by a large agency after the MFA showcase. They had separate print, commercial, theatrical, and even sportscaster departments. They had signed me for commercial representation, which was a good way of testing the waters to see if I could book jobs. I had stayed active in the theater after UCLA, joining a company led by our voice and speech instructor. I'd starred as Andromache in Euripides's *The Trojan Women*, keening as my character said goodbye to her young son, who was to be thrown off a cliff, one of the most heart-wrenching moments in Greek tragedy. Salome had attended a performance and came up to me after the show to tell me that I really "went there," which meant that I'd nailed it. Yes!

I'd been told to look for ways to stay in my agent's thoughts without hounding her, so that she would think of me (preferably fondly) when new casting notices came out and hopefully send me out on auditions. One good way to do this was to "drop by" looking great, with a postcard for an upcoming theater show I was in. I was to put the postcard in her hands and exit quickly, as if I was too busy to stay to talk. In other words, I needed to look like I had a job. That's what would make my agent want to get me a job.

I had a show coming up, so I dressed up in a cute outfit and brought my postcard to the agent's office. I was casual yet confident, and she seemed responsive to the postcard. As I went to leave, saying something along the lines of "Off to a meeting, great to see you!," Madelyn—my agent—a short white woman with close-cropped hair, asked me if I'd conditioned my hair lately.

"It looks dry. You might want to try a deep-conditioning treatment."

I thanked her, played it off, and got the hell out of there. In the parking lot I sat in my car, stunned and confused. I'd spent a long time standing in front of the mirror making sure I looked my best for my postcard drive-by. It had been a hot day. But I was always a big moisturizer when it came to my hair. I'd thought it looked great. How could I have been so wrong? Or was *dry* the word that my white female agent used to describe *Black* hair? I decided that it was more likely that I was wrong and needed to take better care of my hair than that my agent was racist. It was easier that way. I didn't know then how my past experience at Madison had informed that choice, and how very much harder in the long run it would be to make it, over and over again.

I never "dropped by" my agent again.

It suddenly hit me that I was no longer immersed in an environment where my academic performance alone would get me from one stage to the next. One thing about the educational system in the United States was that it prided itself on sequential learning—milestones, benchmarks. To do more in a given field required the acquisition of additional knowledge and experience in that field. Once a person had obtained what was required, the gates opened a correspondingly wider amount to allow that person to do more in their chosen field—at least, in theory. There was still systemic racism to contend with in the world of the "Ivory Tower." There was university politics. But on paper, the rules of getting ahead in the game focused on what was inside a person's brain. No one cared if I conditioned my hair in the morning. No one cared if I didn't.

I was now navigating a world where even literally getting in the door of an audition room was a struggle. And how I looked on the outside, rather than who I was on the inside, meant almost everything. When I was invited to try to secure a place at the table of paid actors, I had at most thirty seconds to project some quality in an audition, such as "affluence" in the case of credit card commercials, or "casual fun" for beer commercials. What I could show them in those thirty seconds dictated whether I would be allowed to show them what I could do on a full shoot for $500 a day, and possibly thousands in residuals later. Along with most of Los Angeles, I just wanted my shot. But aside from mostly unpaid theater and one modeling print gig, I wasn't getting anywhere.

The debating nature of my relationship with Steven intensified into frequent arguments. We argued in the morning. We argued at night. I had heard somewhere that you should never go to bed angry, so our arguments lasted until 3 a.m. I took everything personally and couldn't let anything go. I reacted to whatever Steve said, and we both rode those reactions off into the sunset. I left him sitting in restaurants in the middle of the main course. I opened the door of moving cars he was driving and ran away, hoping he'd run after me. That's what happened in relationships that were filled with passion. We argued up and down the Pacific Coast Highway, on the way to quaint little bed-and-breakfasts in sleepy seaside towns. We argued while his sister was visiting, with her standing by in his apartment while we yelled into the receivers of our phones. We clung to the spirits in each other, knowing that when it was good it was really good, and we always tried to get back to that place again. We'd been apart the night before the September 11 terrorist attacks, after some heated argument that meant Steve was not sleeping over. I called him as I watched the towers fall and asked him to come over. For a moment in time we knew what was important, and behaved accordingly.

There were bills to pay. And I was using money I didn't have to pay them. Every month I played a game. The game was called "Which bill can I finagle my way out of paying this month?" I got a rush out of stretching twenty bucks over two weeks. I put $3 worth of gas in my car at a time. I knew that would buy me just enough to get around town and leave me what I needed for parking meters. Then it would all blow up in a sensational backlash of deprivation: I would pull out my credit card and charge a day at the Thibiant spa in Beverly Hills, or a set of makeup brushes at Barneys. It didn't matter if the ladies behind the counters didn't think I could afford it. My platinum Mastercard (that I'd lied about my salary to get) said otherwise. And since I had used the credit cards anyway, I might as well keep using them—for groceries, dinners out, full tanks of gas. It was no use drawing a line. I could always justify crossing it. And the high that I got from it, from the delirious fog that came from being able to buy anything that I wanted, no matter what it was, without even pausing to decide if it was something I *really* wanted, was tactile. I was free, with no limits. My skin felt more sensitive, my eyes sharper. The sky looked bluer. And for the next twenty-four hours after I bought something, I was happy. Then I could go buy something else.

Except, when the credit card bills came, the minimum payments kept getting larger and larger. Sometimes the card companies would grant me a "payment holiday" when I called them. It meant I could skip the minimum payment that month. I thought that was fantastic, that I'd scored something big out of them. But interest kept accruing.

I had to get a job.

The costume designer at UCLA got me a job at the LA County Fair, along with my former classmate Robert, a beautiful, slim white man with brown hair who danced like Patrick Swayze and Mikhail Baryshnikov combined. We worked the Egyptian Tent. We put our MFAs to use walking around as Queen Nefertiti and an unnamed pharaoh, greeting the public and being regal. All I remember was that the days were long and hot. The fair was in Pomona, about a twenty-minute drive inland from LA, in early September. It was a three-week gig of fourteen-hour days with no day off. Robert and I took turns lying down on a cot in an upstairs turret of the tent. But I was earning.

I called my friend Susan to tell her how much I wanted the new Frou Frou CD. It was $12. I thought about putting it on my credit card. I wasn't sure I could afford it. I needed a lot more than $3 of gas to make it to Pomona and back to LA every day. Susan had started taking better care of herself financially and mentioned something to me about balance, between things that were needs and things that were wants. I didn't know about that. I knew one without the other made me feel deprived, and I usually lashed out with a credit card tantrum blowup. The CD was definitely a want, rather than a need. But this time, I'd taken the time to figure that out. I hadn't simply bought it without thinking. I had even given myself time to check my bank account balance. That meant I was actually interested in buying this item with my own money instead of the credit card company's. I'd thought about whether I could afford to spend $12 on something extra, and what it might mean I would have to forgo in exchange if I did. This was new. A few days later I called her with Frou Frou playing in the background on my way to Pomona. Her cheers, and that CD, kept me going through the rest of those twenty-one days. Then on day twenty-two, I went back to being unemployed again.

The first temp job I took was doing data entry in a cubicle in a high-rise office building in downtown LA. After a week of typing long sequences of numbers from a stack of ledgers into computer spreadsheets, I thought I might jump off the top of that high-rise building on my lunch hour. I begged the temp agency to give me something else. A music publishing company put in a request for several long-term temps. I was one of them. The company wanted all of its publishing contracts pulled out of filing cabinets and scanned. I sat in a room pulling staples out of each contract with two other temps, one loudmouthed white boy with eternal facial stubble and layers of skin and muscle and fat spilling over his waistband, and a dark ebony-colored woman who was also large, with long, braided hair extensions and skin so smooth it didn't look real. Her name was Debbie, and she was from South LA. She loved Red Lobster. Like, *loved* it. We would sit in that room and talk about her latest trip to Red Lobster, and about how there's no need to punch through the middle of a toilet seat cover when in a public bathroom stall, because as she said, "Your pee will break it." She was adamant about this. It meant a lot to her that we understood that the extra work just wasn't necessary.

I wandered through bookstores during this time, searching the shelves for something that would calm my big feelings of being out of control of my destiny. It seemed as if, separated from the warmth and comfort of the educational system—instructors, classes, rehearsals, the camaraderie of shared goals and dreams with my fellow classmates—I was now a small piece of kelp cut loose from the larger organism, floating alone in a vast sea to tip and sway with the ocean currents. I possessed no tail or arms, no means of propulsion or intentional movement of any kind. That I couldn't simply catapult myself across the water to the island of regular employment as an actor after completing my training was infuriating. I wanted someone to tell me what to do.

I picked up a book by Natalie Goldberg, *Wild Mind*. The title seemed fitting for me. I took it home and read it like drinking water. I let it pour in. At the end of the book there was a note about how Natalie conducted workshops in northern New Mexico, in a town called Taos. It said that she offered scholarships to people of color. They could write to her about it. And she gave a PO Box address. I pulled out a sheet of paper and a pen.

I called up Kara, a friend who was a couple of years behind me in acting school at UCLA, and asked her if she wanted to start writing together. We could meet in cafés around LA. She'd let me know that she admired some of the pieces I'd done when I was at UCLA. Maybe that's why I'd felt bold enough to reach out. We didn't know each other well yet. But I knew she liked to write.

We'd meet at Literati Cafe, or Urth Caffé, or the Bodhi Tree back when they had a café on the side. We would order some fabulous tea they had, and as Natalie told us to in the book, we'd give each other writing prompts, set a timer, and go. "*I am thinking of*—ten minutes—go!" "*I am not thinking of*—ten minutes—go!" "*I'll never forget*—twenty minutes—go!" When the timer went off and we'd finished the sentences coming out of the tips of our pens, we'd read to each other with no comment. If the café was crowded and tightly packed, we read in whispers, hunched over our notebooks, our breath fogging up our mugs and warming the ink on the pages. I kept Natalie's words in my ear as I wrote, keeping my hand moving, going for my first thoughts, trying to keep what I wrote pure and unedited, so that there was no filter between my mind and the page. I tried to pour it all out, pure like an Icelandic spring, straight from the source. Some days my mind droned on in zigzags and never got still. Other days the writing cut to the quick like a straight razor, sharp and clean and bracing.

Memories surfaced from the deep—deeper than the Mariana Trench, the deepest oceanic trench on Earth. My mind went down this far and back, from the glass standing on the table in front of me with the jagged trapezoid of ice in it, to a fourth-grade math test, and to the map of freckles on the face of the red-haired boy who was my class rival in math in Victoria, British Columbia, Canada, in 1984. His freckles spiraled around his face, and every time I looked at him they seemed to have moved, like a colony of ants. When I scored higher than he did on that math test, I was happy and sad at the same time. I had lost my competition.

The act of writing electrified me. All of my pain, my confusion, my discouragement, my uncertainty about where my career was going or if I would ever have one, had a place to go: it went to the page. I remembered writing my first poem when I was eight. I titled it "Watercolor." When I wrote it—something that came from me that didn't exist before, and now did—I felt alive, like dancing light and breath. Like Ethan Hawke's character in *Dead Poets Society*, when Robin Williams helps him finally let out, through a poem, what's been locked up inside of him for his whole life, and the world shifts. I had found that feeling again, and suddenly it didn't matter so much that I had no idea what the hell I was supposed to do with my life.

Natalie wrote back and gave me a scholarship to offset the costs of attending one of her retreats in Taos. I took a plane from LA to Albuquerque, then a slow shuttle that snaked through the Rio Grande Gorge, through Santa Fe, and up into the hippie ski town of Taos. It was my first time this far north in New Mexico. I had fallen in love with the Southwest during my time in Flagstaff, Arizona, at Lowell. Not long after Steven and I had started dating, I took him on a road trip to some of my favorite spots in this part of the country: the river-carved chasms of the Grand Canyon, the sandstone Mitten Buttes of Monument Valley—all of the places where I had felt free and desired nothing. We'd hiked up one of the buttes and stood on the rocks to look across miles of open, sparsely populated plains in reds, browns, and golden tans.

Coming up the gravelly path to Mabel Dodge Luhan House, I knew I was in for something special. The smooth, reddish-brown adobe seemed familiar to something ancient inside me. The smell of sage and piñon, the dry crispness, and the cold air of winter wrapped themselves around me. I dove into the structure laid out before me. Wake at 6:30 a.m. Sit zazen, followed by breakfast—a bounty of egg dishes, fruit, muffins, hot cereals. Grab a to-go mug of hot tea and sit on a chair outside before morning class at 9 a.m. In class, sitting meditation, walking meditation, and writing, in large groups and small groups. Lunch, nap, open time to wander into town to browse the stores, stare at paintings, write. After dinner, evening class to sit, walk, write. Then the slow walk to bed in Ansel Adams's room or Georgia O'Keeffe's or D. H. Lawrence's. They had all stayed here. Their spirits swept through the rooms with the New Mexico wind.

I wrote it all in Taos. How I'd loved the stars since I was a child, why I left astronomy, where I was now, what I loved about acting. I asked myself the questions that had toured all of the cells in my body for years. As it all poured onto the page, some things became clear. The two loves were true loves. They each held an entry point and an ignition source. It was no use to try to make one wrong or any less "me" than the other. Each could trace its own path from my head to my heart. No more proof would be requested. I couldn't yet make sense of it all, or even accept it all in me yet, without question. I was beginning to see that the path I was meant to follow would require that of me eventually.

But things were becoming clearer. I believed in the beauty of the universe, and in the power of a story well told through its characters. That was the common thread: story. Onstage, the story's characters talked, growled, breathed air. The story of the universe had its own cast of characters—planets, stars, galaxies, light, distance. These characters lived and had fates too.

I returned to Taos six months later, this time for a silent retreat, where we spoke only when reading our writing aloud. Afraid at first of being swallowed whole by all that silence, I reveled in it. I never wanted to talk again, or hear anyone else speak. So much energy out when the mouth opens. In silence I could go nowhere but inward, and I had to find my answers there, rather than from outside of myself. It was an important lesson for me. How often had I looked elsewhere for a clue to my next move, for permission, for acknowledgment and approval? Do you know what I mean?

Kara told me about a town an hour and fifteen minutes' drive from LA called Ojai. I found my second place of renewal at the Ojai retreat, atop a hill in the Ojai Valley. I brought a few days' worth of food and slept, wrote, painted badly, and walked slowly down the stone-lined paths around the retreat. I sat and stared into the valley basin below and at the surrounding mountains, letting my eyes lose focus as my mind sailed back into memory. My body began to know my mind, and the two enjoyed each other's company.

Some questions remained. What was I to do about these two loves, astronomy and acting? Where did I fit?

Break

I was speeding around the snaking curves of Mulholland Drive in the hills far above the city, on my way home from my latest job as an outreach and communications assistant at a prominent cultural center in Los Angeles. My cell phone rang. I answered, controlling the steering wheel with my left hand and easing off the gas slightly with my right foot as I kept up with the changing driving angles. It was Julie, the producer of the movie I'd auditioned for a few days earlier. Her husband was an instructor at Exeter, and I'd activated that connection when I'd found out about the project, getting myself in for an audition after seeing him at an alumni dinner.

"Want to be in *Nine Lives?*" she asked.

Oh my God!

After my initial squeals of delight, which she entertained graciously, she mentioned something about the union rate. I told her I didn't have my Screen Actors Guild card yet. Having a SAG card would mean I could audition for mainstream shows and films, and get paid higher rates than I would for nonunion gigs. A SAG card was the entrance ticket into the acting big leagues. There were also guidelines for how long directors could have union actors work on set before they were required to give them breaks, and when overtime pay kicked in. Every actor who didn't have their SAG card wanted one.

It was something of a Mount Everest climb to actually acquire a SAG card. You either had to be an extra (a background actor) on a union gig for at least three shoot days, or you could join after being in another union for one year, like the theater-specific Actors' Equity Association. Or you could be "Taft-Hartleyed," which meant that the union gig could hire you, the nonunion actor, if they thought you possessed a skill that couldn't be found within the union talent. This is the preferred way to become SAG eligible. It doesn't require nearly as much work for the actor, but it requires that the producer on a film judge the nonunion actor to be worth the time to do the extra paperwork.

"I guess we could go ahead and Taft-Hartley you."

My mouth hung open on the phone. I'd made it.

Nine Lives was directed by Rodrigo García, the son of the world-famous author Gabriel García Márquez. The movie shows about fifteen minutes in the lives of nine different women, some of whose lives intersect. All of the women are struggling in some way—grappling with a decision, resisting their circumstances, at a crossroads. All are "looking to live," as Salome used to tell us. No one wants to see a movie about women who are wallowing in their pitiful lives. But a movie about women who are going through hell and not taking it lying down—maybe. Perhaps even more interesting, the movie was shot with a Steadicam. Each of the nine scenes constitutes a single take—there are no internal cuts in a scene. The camera follows the entire length of the action for the full fifteen minutes.

I was not one of the nine women. Those roles were played by major actresses, like Glenn Close, Sissy Spacek, and Robin Wright. I was a prison guard in the very first scene, which focused on Elpidia Carrillo's character, a woman in prison who is just trying to have a conversation with her daughter on the phone through a glass wall, and can't. It's a heartbreaking scene, one I would not truly understand the weight of until much later in my life.

I had fourteen lines in the scene and none of them were cut—that's a big deal for an actor. The "no internal cuts" rule was on my side. We filmed in a section of an actual prison outside Los Angeles. My scene partner was Miguel Sandoval. Growing up, I'd watched him in *Clear and Present Danger*, one of my favorite Jack Ryan movies, with Harrison Ford. To act alongside him now felt surreal.

There was a wrap party, with Dakota Fanning, another one of the cast members. There was a premiere, at the Sundance Film Festival. Steven and I flew to Park City, Utah, and stayed with his sister Lori and her family. We were too green to adhere to the informal dress code of "ski chic" like all the celebrities. We sat at a long table at the private dinner near Jason Isaacs, I in shiny black embroidered slacks and a long paisley patterned jacket, Steven in a long-sleeve dress shirt and pants. Neither of us could believe we were there. In the theater after the screening, I filed up onstage with Glenn Close and Sydney Tamiia Poitier (daughter of the Oscar-winning Sidney Poitier) to bow and be applauded. Steven sat in the audience and clapped, then turned to his left to see Steve Buscemi give a slow bobbing nod of approval as if to congratulate him on his choice of wife.

The night was long and crowded as we snaked our way through packed clubs full of Sundance attendees in thick parkas and snow boots. As we left the last party and stepped out into the street holding hands, a feeling of relief washed over me. This was all so exciting and wonderful. And yet somehow there was a foreignness to it. I couldn't tell if it was because it was all new and I wasn't used to being allowed into elite acting circles the way I had been for the past months. Did I belong here? Had I found my people? Or was this as strange a world as the astrophysics department at Madison had been? I didn't know the answer. As I stepped over a pile of hard snow my head craned back for a split second, and something flashed in front of the right corner of my eye. It was bright, twinkling high in the crisp black night. It was a star.

Recovery

Sometimes you can wander around for years, and then in an instant, true north stands above the tree canopy beckoning. There is no struggle, no extraneous thought. The path is made clear.

That one look back up to the night sky sent a silent shockwave through my body that wouldn't be forgotten. As I inched through bumper-to-bumper traffic in LA, to more auditions, the LA red carpet premiere of *Nine Lives*, a commercial acting class, a new neighborhood to put flyers up in for the museum programs, that memory reverberated through my tissues. Of all the characters I played during my years as an acting student and struggling actor, with occasional successes and small movements forward, I discovered that the character I was most interested in embodying was myself. And that character—me—was someone who missed the night sky. I looked up again. I could make out a couple of stars through the gray nights of a smoggy city. Their lights strained through the haze, pulsing lighthouses in the foggy sea air. It was enough to wake me up.

I was sick of living paycheck to paycheck, deciding between paying a bill or putting gas in my car. I had finally answered for myself that question that had been rolling around in my head for years now: Did I miss astronomy? The answer was yes. Maybe I could get a better-paying day job studying the universe than my underearning cultural arts center outreach job.

I emailed my old research advisor from Lowell and my undergraduate days, Deidre Hunter. She told me that jobs were often posted on the website for the Infrared Processing and Analysis Center (IPAC) at Caltech. So that same day I went to the website, and there it was: an ad for a help desk operator working in support of the Spitzer Space Telescope. The person they were looking for needed to have a bachelor's degree in the physical sciences, a few years of professional experience, and good oral and communication skills. I knew it was me that they were looking for. I knew it like I knew that pi is equal to the circumference of a circle divided by its diameter.

The Spitzer Space Telescope was one of NASA's Great Observatories. Like the Hubble Space Telescope, Spitzer was in space. But Spitzer looked at the universe with infrared eyes, instead of visible eyes. It saw things that humans can't see—dust around young stars, stars still being born. Those of us who worked in support of this mission called Spitzer "the Spacecraft."

I never mentioned in the interview that I was an actor. I didn't want to be labeled as a flake, which is how people in LA who weren't actors thought of people in LA who were. It was right there on my résumé, but I didn't bring it up. Once my supervisor on the Science User Support Team was confident that I was dependable, leaving over my lunch break to go on auditions wasn't a big deal. But I ended up wanting to stick around and go to astronomy talks. Since I'd left the field, the discovery of planets orbiting stars other than the Sun—called extrasolar planets, or "exoplanets" for short—had exploded. Scientists came to IPAC to talk about all sorts of astronomical phenomena—galaxies, the interstellar medium (the space and dust between stars), supernovae (exploding massive stars), black holes (what can happen to the most massive stars after they go supernova). But it was talks about these new exoplanets that kept stopping my heart.

Word got around that I was an actor, and someone at work forwarded me a posting about a science TV show. PBS and *Wired* magazine had partnered to create a news magazine–style program about science. They were looking for scientists to host the show, so naturally they contacted Caltech. I sent in my résumé and headshot that day. I was contacted by a production assistant within hours. "We have to get you in to talk to the producers today."

I was meant for that show. My science background, my acting training. It was inevitable. I hadn't been able to see it. Even while I had accepted the fact that both worlds coexisted inside of me, I couldn't fathom a scenario that would allow me to inhabit both worlds—the worlds of science and theater—out in the world, as a career. Nevertheless, a scenario had presented itself.

Wired Science shot its pilot episode at Yellowstone National Park. I walked around steaming hot springs with microbiologists, interviewing them about acidic environments and the extremophiles that called these hot springs home. I dipped measuring instruments into the mud pots and traipsed along the rocks and grassy mounds in my hiking boots, jeans, and vest. I watched buffalo stop in the middle of the road and stare at the long line of vans and cars waiting. Here, the animals dictated the timetable.

Our show was greenlit for more episodes. I was ecstatic. Again, for real this time, I thought that I had "made it." I was going to be the regular cohost of a science TV show. When my agent called to tell me that the producers had decided to go with a different host instead of me, the world got very small. I was standing in the kitchen. Steven came in. Without hearing the phone conversation, he knew what had happened. He was in this business too.

It wasn't personal. There was money involved, and they had to go with the person they thought would read best to audiences.

That was all bullshit. I was devastated. *Those motherfuckers.* How could they? To get so close, and fail like this. I collapsed in Steven's arms and sobbed until there were no more tears left. There went the brief dream of a life of both.

☽

I decided to apply to the astronaut candidate program. This is what had drawn me to the stars in the first place. No use dancing around the goal. I would go at it head-on. I had a well-paying job; I made enough money. I hired a physical trainer to whip me into the kind of shape I needed to be in to be a viable candidate. I needed to be able to jog at a comfortable pace without gasping for air, to be put on a treadmill for a fitness test and pass with my heart rate in a normal range for my height, weight, and age.

I also decided to get laser eye surgery. For the first year ever, NASA was allowing applicants with uncorrected poor eyesight to be eligible for LASIK or PRK procedures. My corneas were thin, so I had to get the old-school PRK, like the early test pilots whose eyes had to withstand high g-forces. Steven sat in a viewing room watching the surgeons operate on my eyeballs, propped open to prevent me from blinking during the procedure. I wore thick white eye patches for two weeks and stayed home listening to the sounds of the television as I sat on the couch looking like a large fly.

My life was simple. Wake up at 5:30 a.m. Dress quickly in sweats, grab a bag full of work clothes and makeup that I packed the night before, kiss husband, and head out. Drive to Pasadena via the 101 to the 110 to the 134 to the 210 Freeway. Exit Lake Avenue. Park in the lot next to the gym. Do treadmill or elliptical, weights, circuit training designed by my trainer Mike. If I was meeting him, he took me through it, and we went on to the track to run. I watched my muscles become more defined. I watched my face get thinner, felt my watch band loosen on my right wrist.

I submitted my application to NASA. And then I waited.

A postcard came months later. I held the thin card in my hand, flipping it back and forth over and over, looking for more information than what was there. It couldn't all have come down to this one small piece of cardstock with a form statement telling me I had not been selected to continue on to the next application round, thanking me for my interest in the NASA astronaut candidate program, and reminding me of how many thousands of other people had the same dream I had. Maybe for as long, or longer. In my mind, no one could have wanted it more.

With a bachelor of science degree, I had barely met the qualifications for the astronaut candidate program. I had no PhD. I knew of no mission specialist—the scientist astronauts who don't pilot or command the shuttle—who didn't have a PhD. I had known it was a long shot, but that didn't help much to remember. All I could think about was how I'd let someone operate on my eyeballs to bring them to the degree of precision worthy of the trust of the United States space program. But NASA didn't care about what I'd undergone. They may not have even had the time to notice it amid all of the thousands of applications. It was a choice I had made, and no one was going to give me anything for it. At least I no longer needed to worry about falling asleep with my contacts on and waking up to them dried out and stuck to my eyeballs.

I focused on what was in front of me—work. I was making more money, and I started doing with it what I saw friends of mine do with theirs—saving it, opening retirement accounts, creating vacation funds, with a trip to Italy at the end of the year as a goal. I discovered that when I spent within my means, including giving myself things I wanted, like monthly massages and flowers, I didn't feel deprived or resentful or impatient. The pull of the credit card became weaker and weaker, the way the surface ocean currents driven by winds dissipate with depth beneath the waves. Eventually, I didn't feel it anymore. There was enough money for it all, to live in the present while cleaning up my past, planning for the future, and being kind to my older self by saving for her, like an older friend told me to long ago. I concentrated on being a more loving partner—one who didn't take everything so personally or so heavily. I let go of some of the weight I had been carrying on my shoulders since I was a child, when I wanted so much to help my mom be happy and couldn't, though I kept trying. I started to lighten up. I let her go, and in the process let go myself. I started to feel my age again. Not fifty-eight or forty-seven. Thirty-two years old.

Pluto

I remember when Pluto got demoted from its status as a planet to something lesser. It was sad, but necessary. It wasn't long after I'd taken the Caltech job. Mike Brown, a planetary scientist whose office was across the street from my building, in the Department of Geological and Planetary Sciences, had discovered something new in space. The object, which he eventually named Eris, after the goddess of discord, was larger than Pluto. It was spherical, not like the chunks of rock we call asteroids—oddly shaped, broken fragments left over from the formation of the solar system. The crumbs on the plate after all of the cookies had been eaten. No—enough pieces had grouped together to form a ball-shaped object called Eris. It sat suspended at the outer edge of the solar system, until finally someone had noticed it was there.

Then something else was found—another object, called Ceres. It, too, was large enough, at about 500 kilometers in diameter, to have molded itself into a spherical shape under its own gravity. Both of these objects were in Pluto's class—small, spherical bodies of rock, a little icy, and not quite by themselves. Pluto's largest moon, Charon, was half the size of Pluto itself, suggesting a more balanced relationship than is usually the case between a planet and its moons. Had Pluto or Charon at one time been something else entirely?

Pluto had always stood out. Besides having an unusually large moon for its size, Pluto's orbit was highly elongated and inclined. You could just about lay the orbits of the other eight planets in the solar system on a plate and they would lie flat. Pluto's orbit would cut through them at an angle of more than 17 degrees. Clyde Tombaugh, a slender young astronomer at Lowell Observatory, searched photographic plates for Pluto. Astronomers had inferred its presence from the motion of the other planets whose gravity it influenced. And finally, one day Clyde found it by comparing plates, noticing its movement between the frames relative to the more fixed background stars.

More and more objects kept getting found in the same class as Pluto, in the outer solar system. It came down to either adding to the number of existing objects that were considered planets, or taking Pluto out. In 2006, the International Astronomical Union (IAU) met in Prague in the Czech Republic and voted. Seas of old white men held up yellow cards, and a few white women did too. I wasn't there. Astronomy was still only a day job for me to support myself while I looked for acting work. I had dipped my toe back in, but only my toe, working on the periphery for the community of people who had actually dedicated their lives to the field. I probably wouldn't have been at the IAU vote if it had happened today. I should join. It sounds official. One thing they do for sure is vote out planets.

And that's what they did that day in 2006. They demoted Pluto from planet to "dwarf planet." They also ratified the concrete and unchanging definition of a planet, for posterity. A celestial body is a planet if it (1) orbits a star, (2) has obtained sufficient mass to attain a spherical shape, and (3) has cleared the area around its orbit—which I suppose means it can have moons but not much else around it. This last part is important. Folks concluded that Pluto had likely been in the Kuiper Belt—a group of thousands of objects orbiting the Sun at the outer edge of the solar system—and been captured from that population. Other objects, like Eris and Ceres and those that were found later, like Haumea and Makemake, also fit this bill of objects lured away from their homes—never fully independent, never completely holding their own as planets orbiting the Sun. So most in the field, and certainly Mike Brown, said it was only a matter of time before Pluto was put in its proper place.

I like that Pluto was out of place. I wish we had let it stay there. The scientist in me says it's a relief to make everything clear and cut-and-dried. We now have rules for planets, and boxes we can check every time a new rocky body is found. But the other part of me that is not logical, not rational, and not particularly measured or consistent likes—no, *loves*—things that don't fit into a box, or even three boxes. I love gray areas and black sheep in families. That part of me, that is all of the messy shades of human, wants to turn a blind eye to Pluto's unintentional transgression. Not only that—I want to give Pluto the book *Playing Big* by Tara Mohr and tell it to advocate for itself, plant its icy, rocky body firmly in the black, and not let go of the label it was given at birth (really, at discovery). None of this "dwarf" business. *Give me my name: Planet.*

Stars and Life

I was promoted from help desk operator—a job where I mostly forwarded the hard questions about observing programs to the PhDs on my team—to scheduler of observations on the Spacecraft as a member of the Observation Planning and Scheduling Team. Scheduling each week on the Spacecraft was a game and a jigsaw puzzle: How efficiently could I schedule my week this time? The less the Spacecraft had to slew large angular distances between targets, the more efficient the week was judged to be. When I got the promotion I was terrified that I would somehow screw up royally and rig the Spacecraft to turn toward the Sun. One opening of its shutter onto the hot disk of our parent star and the Spacecraft would become the Fried, Dead Thing in Space. But that never happened.

All of the things we cherish in this world were made inside the bellies of stars. Stars like our Sun will fade into cool, dense "white dwarfs," no bigger than the Earth but carrying much of the Sun's original mass. Slowly their light will fade over time. The most massive stars are destined to end their lives as supernovae—magnificent explosions of color and energy. In the course of these powerful events, these dying stars expel their contents into space and pass on the products of their nuclear fusion reactions to successive generations of stars. These new stars will form out of clouds of gas and dust laced with these elements from their predecessors. Each generation of stars will become more and more enriched with the remnants of those long gone before them. And planets orbiting these stars will in turn form out of the settling remains of their creation. Life on those planets will develop with the chemical fingerprints of this lineage. This is the ancestral legacy of these stars, and of us. Nothing is forgotten. Everything carries the imprint of a previous life.

In the devastating wake of my rejection by the PBS/*Wired* TV show I had banked my future in Hollywood on, and my rejection by NASA, I answered a help desk ticket from an astronomer. Something inside me turned private investigator, and I decided to google her. I saw that she had a theater background, and I was intrigued. I took a chance, and in my response to her observing question, shared a bit about myself. She replied, "You should meet my friend, Neil deGrasse Tyson. I'll introduce you."

She did introduce us over email. Neil told me that he'd actually seen the pilot of *Wired Science* when it came out. He seemed neither impressed nor critical. When I bared my soul over email, relaying the details of my getting replaced as the show continued with more episodes, he gave it to me straight. He told me that to be considered an expert, I needed a PhD. He said that the Venn diagram of individuals with a PhD who could also talk on television and look halfway attractive in that medium was much smaller than the number of people who had those qualities without that degree. The ingredient that I was missing was what would make people believe me on TV. I could act a lot of things, but I needed the street cred. To get that, I needed to acquire the knowledge. There it was again, staring me in the face, just like the NASA astronaut rejection postcard. Just like the Caltech job, at which I'd reached the pinnacle of how far I could go with only a bachelor's degree. The damned PhD.

Months later, I was standing in front of the Cathedral of Santa Maria del Fiore in Florence, Italy, where I'd journeyed with my husband on New Year's Day in 2008. The trip was luxurious, and paid for with cash. A band in the piazza was playing "Dream On" by Aerosmith. I was waiting for Steven's face to emerge through a small open window at the top of Giotto's Campanile, the bell tower of the Duomo. It was a moment of pure contentment, when I wasn't thinking of what needed to happen next, or what had happened minutes or days before. I hadn't had many of those moments, so it stood out for me. I could have stayed in that square listening to that band and taking in my surroundings forever. Steven's face emerged. I took a picture of him from my distant point 414 steps below. He took a picture from his perch up above. And a single conclusion sank deep into my belly and firmly held me there. I said a little prayer to something somewhere that might be listening and care. *Okay, if this is what I'm supposed to do, I'm willing. I'll apply to graduate schools in astronomy again, damnit.*

I wanted something more for myself than I had. I knew what life was like without a PhD. I was thirty-four years old. I could continue in my current job scheduling observations on the Spacecraft, but I could never go any further at that job without a PhD. I would never have any observations of my own to have scheduled for me, if that's what I wanted. Not without that degree. If I went back to graduate school for astronomy, I would probably be forty when I finished. But, as I'd once heard someone say, fingers crossed, I was going to be forty whether I had a PhD or not. So if I wanted one, I might as well go and get it. I decided to surrender to the direction in which all of my paths appeared to be leading. Now all I needed to do was take the actions, one after the other.

We went to the Basilica of Santa Croce, and I stood in front of Galileo's tomb. Steven took a picture of me there. When I returned home, I began studying for the physics GRE exam. Over the holidays I submerged myself in graduate school applications while reading the Twilight Saga books and playing *Mystery Case Files: MillionHeir* on the Nintendo DS Steven had gotten me for Christmas. I had been trying to learn how to have more fun, and these two pastimes were just what I needed as I dove back into a rebid for a PhD.

Neil Tyson and I finally met in person at a meeting of the American Astronomical Society in Long Beach, California, around the time that I embarked on the graduate application process, Take 2. I was there to help run a booth for Spitzer. Neil and I sat at a table in the brisk winter sun while people walked by and craned their necks to confirm he was who they thought he was. I collapsed into a pile of tears and bared more of my soul. The memories of my year at UW–Madison resurfaced, the plummeting grades, what the old, wispy-haired white male professor said to me. And even further back, the years struggling with physics at MIT, fighting until I was raw and ragged for every decent grade I got, and having to accept with the most bitter taste in my mouth the bad grades that I couldn't seem to turn into anything else. I wanted to come clean to Neil about my track record. Maybe he would confirm what that professor had told me, and I could let go of this nagging dream and get on with choosing another one I was better suited for. Maybe he'd say getting the street cred to be on TV wasn't worth it, that I should just stick with what I was doing and hope for the best.

Instead, Neil told me his own story. His wasn't much smoother than mine had been. Doubt. Struggle. Judgment by others. He'd felt like a fish out of water at times, because in addition to doing science, he also valued talking about science with others who might not otherwise understand. He'd fought hard to stay in the game. Then he'd changed the game by making his own rules. And here he was.

He offered to send a letter on my behalf to the graduate schools I was applying to. It wouldn't substitute for the regular letters of recommendation I'd need from my former professors and advisors. But his, he said, would swoop in later, after the rest, like a little birdie to whisper in the ears of the members of the department admissions committees. Then he pulled a folded paper from his jacket pocket. As I wiped my tear-stained face, he unfolded the paper and read it to me. It was the first draft of the letter. He had written it before we met in person. I don't remember much of it. I was overwhelmed. But I remember it ended with him saying that I had a special combination of characteristics that meant I carried the potential to be something none of their departments had seen before. Something great.

The day I got a call from the chair of the astronomy department at the University of Washington telling me that I'd gotten into grad school, I hung up the phone and looked up at the poster of Carl Sagan that was tacked to the wall in my office. Carl was sitting on one of a sea of planets that had been drawn in the background—the planets he had always known were there. He was smiling down at me, as if to say, "Isn't that something. Welcome back." The quote at the bottom of the poster read, "Somewhere, something incredible is waiting to be known."

Hair

There is a plaque on the wall of my office that shows the bottom half of a woman holding a suitcase at the foot of the steps leading up the stoop of a house. The plaque is painted in shades of terra-cotta and soft greens and browns, and reads, *Sometimes right back where you started from is right where you belong.* This message reminds me that going back to something doesn't always mean moving backward. Sometimes moving forward means returning to the thing you left and thought you'd lost, because, as it turns out, that thing was what lit you up inside, and you finally remembered why.

The promise of returning to something I'd left behind felt new and exciting. All I had were the memories of who I had been when I was last there. But I knew I was a different person now. I carried with me the experiences of acting and being out in the world over the years since I'd left. And hopefully the maturity that came with them. I'd gotten married. I'd learned how to take care of myself in relationships and with my pocketbook. I'd gotten to know who I was and had begun to accept myself. That was still a process. But I'd made a lot of progress. I hoped that would count for something.

In addition to the requirements for life to exist, there is a separate set of requirements for living organisms to move beyond merely surviving into thriving, flourishing without boundaries or limitations. These requirements can vary widely and are sometimes incredibly unique, even surprising, depending on the organism. These could include a certain temperature range found in extreme environments, like the steaming hydrothermal springs of Yellowstone National Park, or the icy crystalline shapes called frost flowers that form in arctic landscapes. Certain life-forms thrive under pressure at the bottom of the ocean, while others proliferate under extremely acidic conditions, like those found in a pool in the Dallol crater in northwest Ethiopia. We call these organisms extremophiles because they don't just tolerate extreme conditions—they thrive in them. They know what they need to feel fully alive, and they actively seek it out.

I wanted to bring all that I had acquired into this next pass at a graduate degree in astronomy. Merely surviving was no longer enough. Back then, my head had barely been above water, and at times it was submerged. I knew that, like an extremophile, there was a particular set of conditions that would allow me to thrive. I had ideas about what some of those might be. But I wasn't sure about all of them. I was hopeful.

Central to this effort was a desire to shed the things that I wanted to leave behind. I wanted to carry no apology anywhere in my body, for my past, for my mistakes. I wanted to leave behind who and what I had tried to be so that other people would think I was beautiful and successful, and embark on this new journey light and true to the person I was born being. One surprising area where this manifested was my hair.

I used to sit in a salon chair for four hours getting a Jheri curl. And then in later years, a Cosmopolitan curl. Then a Leisure curl. Then a Wave Nouveau. This all started at age nine, when I went to live with my stepfather in Victoria, British Columbia, Canada, for a year while my mom stayed in La Jolla with my grandmother to finish her dissertation. My grandmother wouldn't be there to braid my hair, so my mom took me to get the Jheri curl—a body perm for Black hair. The Jheri curl's purpose was to make Black hair curl in a soft, flowy, "white girl hair" way that laid down the back rather than stood up in a big afro.

To get Black hair to do what white girls' hair did naturally required pouring harsh chemicals into it—relaxers to straighten, followed by curl activators to cement the perm. The curl activator came after several hours of slowly rolling small patches of hair onto rollers with tiny, thin strips of paper between the hair and the roller itself. I never asked what the thin strips of paper did. I simply handed strip after strip to the hairdresser as they stuck their hand out to the side, right by my ear, waiting for a new strip. When all of my hair was rolled up, a long cotton strip was wound around my hairline where the rollers at the edges touched my skin. I knew what this was for—to absorb the curl activator that would be poured over next, and to keep it from reaching my skin, where, who knows, it might cause a chemical burn. I remember holding a towel against my eyes to protect them from the solution, should the cotton barrier fail. I remember how cold the solution was as it poured, how my head felt like it was under siege by an army of ice soldiers, or an ice queen, like Elsa from the movie *Frozen*, casting a freezing spell on my head. Section by section, my hair succumbed to the curl, the darkness, the spell. I was under Elsa's control, and there was no going back.

When the Earth was first new, everything was hot and molten. Suspended in the cold of space, heat radiated away, and the Earth cooled down and solidified. All the chemicals present in a rock at that time got locked in. And unless the rock got heated to its melting point again, that's where they stayed. Nothing else that can be done to a rock can change the date on its birth certificate. The rock is as it was when it was born, and will continue to be so for all time. This is how we measure the age of the Earth: by measuring the age of the chemicals inside the oldest rocks we could find—from the Jack Hills of Australia, from the surface of the Moon, and from asteroids in space, whose fragments fall to the Earth as meteorites. This is how we know that the Earth is approximately 4.5 billion years old.

When I was thirty-four years old, in the summer before I returned to graduate school in astronomy, I finally tired of the Wave Nouveaus. I was tired of covering up whatever my hair had been like in its natural state at birth, what I received only brief glimpses of in the form of new growth at my scalp line under the perm before I ran back to the salon. I was tired of having to drench my hair in greasy solutions to get it to look like bouncy, loose, curly hair. I was tired of hating to get it wet—in the ocean, in a pool, in the shower. It took a week of drenching it with hair spray and curl cream to get it out of a frizzy mess and back into white girl territory. I was tired of worrying whether Steven was going to recoil with sticky hands when he touched my hair. I was tired of feeling the area near the back of my head and noticing a distinct step where the hair was thinner. Where had my hair gone? What remained was hair that had survived the friction created by ponytail ties and scrunchies through the years. I was tired of calling salons every two-and-a-half months to make another appointment for a Wave Nouveau and being told, "Oh, we do relaxers and cuts and press and curls, but no Wave Nouveaus. They're too smelly," or "I'm sorry, the stylist who still does those no longer works here." I was tired.

I walked into a high-end salon in the Fairfax District of West Hollywood, a three-minute drive from my apartment at the corner of Laurel Canyon and Crescent Heights. I told the woman at the front desk that I wanted a haircut. A tiny white woman with a cute hazelnut-brown bob framing her face and large doe eyes approached me with a big smile. Her name was Chevonne. This sounded like a Black girl's name. Maybe I had a chance, I thought to myself. I told her I hadn't seen my natural hair in twenty-five years. She led me to her station and put delicate hands gently on my hair as she looked it over. She combed it out, surveyed the large afro before her, then started cutting it dry. She told me that this would help her to cut my hair more evenly than if she tried to cut strands that might curl up by different amounts when wet. She shaped, gently extending sections, balancing comb and clippers in one hand, then passing the clippers back and forth between hands, and combing out again. She led me to the sink to wash, then back to the chair, where she applied some light curl cream. When she was done, she showed me in the big mirror before us, and gave me a handheld mirror to use to see the back. I remember her saying, "You have beautiful baby curls." And indeed I did. Somewhere between tight and loose, coarse and fine, were me and my curls. My short, dark-brown hair was soft, shiny, slightly thicker in the back than on the top and sides, and oh so chic. Chevonne, the white woman with the Black girl's name, had uncovered—recovered—my original hair. I truly felt like I had been reborn.

I didn't know it then, but there was more that I should have shed. More that I needed to shed to be successful, to just plain make it easier on myself. But you have to start somewhere. And the rest you learn as you go.

RETURNING

Solar Cycle

The Sun has a cycle that is eleven years long. This eleven-year cycle marks changes in the Sun's magnetic activity and appearance, which affect things like the Sun's radiation output, the number of sunspots—dark freckles on the surface of the Sun—that are produced, the number of flares, and how much material from the Sun is ejected through space in large loops of plasma called solar prominences. After eleven years of relatively quiet, uneventful existence, the Sun outpours in one large tantrum everything it has been holding in for the last eleven years. The Earth bears witness to these outbursts. Electric grids sputter. Sometimes they overload completely, causing power outages and mass inconvenience.

In the fall of 2009, at the age of thirty-four, I entered a graduate program in astronomy at the University of Washington in Seattle. It had been eleven years since I was last a graduate student in astronomy. Coming back was like entering another country in whose language I'd once been fluent but was now foreign to me, the sounds awkward and strange in my mouth. I had been cheating on astronomy with acting for over a decade. I wondered if astronomy would take me back.

Nothing can prepare you for graduate school. Not even MIT. Between life as a college undergraduate and life as a graduate student, there exists the Grand Canyon of chasms. There was an enormous amount of work, yes. That is the same in both worlds. But it's different when you become a graduate student. Suddenly it isn't about punching the clock of hours needed to finish those monstrous problem sets and papers—an all-nighter here and there—and once they're done, you're home free with nothing to do. Maybe start the next problem set if you're conscientious. In undergrad, your time is your own, to read, dance, hang out, have sex, take a train to Quincy Market to get a bag of cookies from Mrs. Fields or a lobster roll from Legal Sea Foods. My roommates and I saw movies. *Jurassic Park.* We sat in the front row of the theater during *The Shawshank Redemption,* our necks craned like ostriches, hanging on every word of Morgan Freeman's rich, deep voice. I was in love with that movie.

In graduate school, there is no end to work. There is no one who will tell you, "Great job on that problem set! Take a few days off and relax." There are no vacations. You take them intentionally, consciously, and in full "screw this" mode to protect your sanity. In graduate school, the first two years are taken up mainly with coursework. You got the grades in college and did enough research to lead the admissions committee to the conclusion that you had the potential to become some kind of reputable scientist. But once you arrive, the slate is wiped clean. You begin again.

Courses. I would take thirteen of them over my first two years. Observational Astronomy. The Interstellar Medium. Galactic Astronomy. Extragalactic Astronomy. The Solar System. Magnetohydrodynamics. Statistical Thermodynamics. The list went on. And I would be going for a dual PhD in astrobiology along with my primary field of astronomy. The University of Washington is one of only a few institutions in the world that offers a dual-title PhD in astronomy and astrobiology, and I knew what I wanted to study now. It wasn't like before, when I let myself be carried with the current into graduate school. I had a well-paying job, a husband with his own well-paying day job, and a life in LA, with our favorite movie theater and Ethiopian restaurant. *God*, that food. I could eat there every day and be happy.

To leave all that, to pick up and move myself and my husband to a place where there was no job waiting for him and the Sun only shines in the summer—I had to want this badly. I was not there to dick around, wondering as I studied, *What kind of research do I want to do . . . ?* None of that. I knew what got me going, and it was planets. Not just planets, but a specific kind of planet—small, rocky ones, like the Earth, that might have oceans and lakes, and maybe life. Astrobiology is the study of the origin, evolution, and distribution of life in the universe. What does life need to survive? What types of planets are most likely to support life? How do we figure that out? And where else in the universe does life exist? I wanted to be able to answer these questions. So there were courses to take in astrobiology. There was a research rotation to do in a field that was as far from my chosen discipline as I could get and still be doing science. At the end of all of the coursework, I would take a six-hour-long exam to prove that I had stayed awake during those first two years, that I had demonstrated some mastery of the fundamental knowledge comprising the field of astronomy. If I passed it, I would be cleared to stay and do research, and eventually get a PhD. If I didn't pass, well . . . there were other options.

Before I left my day job at Caltech to return to graduate school, I canvassed the PhDs in my group. They were all around me. I was one of a handful of people working for Spitzer who didn't have a doctorate. I wanted to know anything, everything that might help me get out of graduate school alive, with my degree and my sanity intact. I asked a guy named Kartik if he had any suggestions. He told me something about a three-legged stool, how research and coursework and teaching are each their own leg on this stool. At any given time during my graduate career, one of these legs was probably going to fold, its strength and effectiveness diminished. But it wouldn't always be the same leg. The trick to keeping the stool—I interpreted this to be a metaphor for my livelihood, hopes, and dreams—upright was to stabilize the other two legs so they could take up the slack of the weaker leg. And to reinforce that stool with a personal life.

This is what the Kepler space telescope hoped to do when its reactor wheels failed. NASA's *Kepler* Mission launched in 2009, the same year I started graduate school at UW. Its sole purpose was to stare at one patch of sky and measure the dips in brightness of 150,000 stars, caused by planets passing in front of them. In 2012, Kepler lost one of its reaction wheels. A reaction wheel is helpful for keeping a telescope pointed at a star. Losing one would have been okay; Kepler had three more. But in 2013 it lost another reaction wheel. Four years into the mission, this was not good. It was time for Plan B. And the K2 mission was born.

K2 was a repurposing of Kepler. A new approach. It was a more limited arrangement, but a way to use what we had to get what we needed: a better understanding of the number of planets that were out there, and how many were small and maybe rocky like the Earth. These planets could be investigated further by later missions with different goals, like determining what was in these planets' atmospheres, and whether they could harbor life.

K2's fields of view would shift, but all would center around the ecliptic plane, where all the constellations of the zodiac reside. Those are the constellations whose names are attached to your birth date, because each corresponds to the constellation that was in the Sun's background path as the Earth traveled around it on your birthday. Their horoscopes appear in newspapers or online, if you're into that sort of thing. (Don't tell anyone, but I love to read my horoscope, even though I know it means nothing. Astronomy is a science. Astrology is just plain fun.)

K2 could operate fine using thrusters and the two remaining reaction wheels. The second-chance afterlife of Kepler, K2 would extend the mission to 2018 and the number of planets discovered by this mission to more than 2,600, over nine years. Many of these planets are promising candidates for habitable worlds.

☾

When you return to something after a long time away, the breath in your lungs feels different. The air is strange in the room, unfamiliar. Light falls on surfaces with a slightly different hue.

It was winter 2010, March maybe. It was my second quarter in graduate school. I had had a quarter to acclimate, during which I sat and listened to other students give talks on scientific journal articles they found interesting in the weekly journal club seminar on Friday afternoons.

First-years weren't required to give a journal club talk the first quarter. We got to sit in the seminar and watch other more senior grad students present the basic facts about the paper they chose and its major conclusions, and pull a coherent conclusion together about the significance of the paper in the broader context of the subfield. I'd had a chance to hear close to twenty people give talks. Then it was my turn.

Steven and I were living in a ground-floor triplex apartment on 15th Avenue NE in Seattle. It was about a forty-minute walk from the door of our second bedroom, which I used as a home office, to my office door at school, which I shared with a sixth-year and a seventh-year grad student, Jillian and Amy. I hadn't started walking to school yet. That would come later, after the borderline high blood pressure diagnosis and my doctor telling me that it wasn't a good idea to have a linear increase in my weight into forever. In winter 2010, my only concern was getting school right—getting good grades, or at least good enough not to raise flags and call attention to myself.

I had learned that I could never be late to class. There were thirteen students in each class: me and the other seven classmates in my year, and the five members of the second-year class. Classes were offered in alternate years, so first- and second-years took them together. But out of those thirteen, the only African American—male or female, teacher or student—was me. If I was not in class, or if I was late to class, it was noticed. The following year, a Black man arrived in the next graduate student class. We shared knowing smiles in the hallways, kindred spirits even with nothing else in common. But until then it was only me.

One day, I was in the home office getting ready to take the bus to school to give my first journal club talk. I had chosen a paper about the two-faced nature of Saturn's moon Iapetus. This moon is strange, and that's why I have always liked it. One side is bright and light, a creamy white. The other side is chocolate brown, spilling into the creamy side. The edge between the two sides isn't sharp. It's jagged, the chocolate seeping into the vanilla. This moon has been a mystery for a long time. What is the cause of its appearance?

I thought of myself, and how others have wondered that about me. *Well, she looks Black, but she doesn't look Black. And she doesn't "act" Black. Where the hell is she from?*

As I sat in the study, I couldn't breathe. I felt hot. My stomach had become a knot, and the bathroom my frequent stop. I tried to tie my shoes, but my fingers wouldn't work right. The laces went in circles and never caught. Steven came down the hall as the tears broke through the corners of my eyes. He kneeled down in front of me, gently nudged my hands to the side, and began tying my shoes.

"I can't do this," I sighed, my voice wobbly and teetering on the trapeze.

"Yes, you can," he told me. "You know how to speak in public. You can talk circles around these wonks. They're gonna be clambering after you to tell them how you do it."

Steve drove me to school. I sat in the passenger seat trying to collect myself. I pulled a compact mirror out of my purse and began fussing with my hair. I searched the whites of my eyes for any trace of pink that might give away the breakdown I'd just had. I reapplied my lipstick, blush. I was not brave enough to put on more mascara at this point, even on only my top lashes. I didn't trust myself not to fall apart again sometime in the middle of my talk.

The journal club talk did not go well.

A few sentences in, someone in the audience stopped me mid-sentence to ask a question.

"In what direction does Iapetus rotate?"

This was a general background question that was certainly reasonable to expect a member of the audience to ask. At the time though, what it seemed like was an extraordinarily rude thing to interrupt the speaker to ask, entirely unrelated to the paper's topic—how Iapetus could have possibly gotten its two-faced-looking appearance.

I knew how to speak in front of people. I was a trained actor! But in the world of the theater, there is something called "the fourth wall"—an invisible boundary between the performers and the people in the audience. The audience was over there, and I was up here, and no one crossed that invisible boundary unless they were invited, as they might be in some kind of call-and-response element of a performance. Now, in the sciences, a field all about inquiry and theories and observations and analyses and scrutiny, scientists could ask questions, not only after a talk, but during, and even before if they felt like it. I hated that.

Of course, I wouldn't have felt nearly this strongly about that person's question if I had known the answer. But I didn't. I hadn't even thought to look up the direction of the moon's rotation, or any other fun facts about this solar system body. And so, I could have said to the audience member, "Gosh, I don't have that information offhand, but I'd be happy to look it up after my talk." Instead, I looked at the individual, then back behind me at my slide showing Iapetus, its bright snowy side butting up against its cocoa-brown side. I then looked back at the audience, raised my arms halfway up at the sides, and turned my body to the side, in a hula-like motion. Then I muttered something along the lines of, "Like this."

Silence.

The air hung thick and hot in front of me. I decided to ignore what I'd just done and proceed with my talk. But the stain of those first few moments left a rancid taste in my mouth that I couldn't forget.

In fact, the direction of Iapetus's rotation was quite relevant to the paper's conclusion—that the chocolate-brown color on Iapetus most likely derived from a dust ring around Saturn. The ring appeared to be caused by ejected material generated by impacts on one of Saturn's other moons, Phoebe. This dust ring was in turn shedding its contents onto tiny Iapetus, painting its surface with the dark material. The darker material is more absorptive than the bright water ice on the moon, so that ice was moving to the poles where temperatures were colder. This dance of dust and ice around the planet is influenced by which hemisphere is leading as the moon rotates relative to the dust ring, and which is trailing. Phoebe and its dust ring rotate clockwise around Saturn. But Iapetus rotates counterclockwise, as viewed from above the moon's north pole. This causes the moon to catch material from Phoebe's dust ring like a shovel scoops up snow. The leading hemisphere receives the bulk of the dust and debris from the disk, and is therefore darker. The trailing hemisphere is where the water ice retreats to keep itself cold enough to stick around. The result is this yin-yang presentation. The slow rotation of Iapetus—it takes seventy-nine days for the moon to turn once on its axis, compared to the Earth's twenty-four hours—means that the dark material absorbs a lot of sunlight. This helps any water ice present within the darker material to boil away into vapor and then solidify again in cooler regions on the moon, further increasing the contrast between bright and dark. I wish I had known all of this then. In my daydreams I rewrite the story of that first journal club talk Q&A, and I know everything.

After journal club was over, back in my office, I commiserated with my officemates. Jillian and Amy were goddesses to me, so far removed from my early, first-year life. They had already proven their right to be there by taking thirteen courses, passing their qualifying exams, then their general exams for doctoral candidacy. They were finishing up their dissertations. They worked wherever they felt they'd be most productive, whether that was in the office or in a coffee shop in downtown Seattle. Jillian would bring her knitting to colloquium and sit in the back row of the lecture hall in her black jeans, boots, purple sweater and hair to match, knitting gloves with the fingers cut out, or hats, or socks. I dreamed of being where they were, of having the confidence to do hobbies in front of an astronomer giving a talk. In colloquium, I strained to understand as much as possible, viewing every interaction with the world I'd been away from for so long as an opportunity to show my worth and prove that I belonged.

Amy and Jillian told me that the journal club talk wasn't nearly as bad as I thought. Amy showed me a quick way to blow up figures while maintaining the resolution of the image, something else I'd neglected to do. They both praised how I had spoken to the audience. "I wish I could speak like you do when I give talks," one of them said. This was an incredible compliment. I did have an advantage over the other students. I knew how to speak in public. And I practiced so much, like my life depended on it, that my facts sounded polished. I would be asked multiple times during those five years of graduate school if I memorized my talks. The truth was that I practiced and rehearsed so many times, that I pretty much knew what I was going to say by heart.

But I became a deer in the headlights in the face of questions that flew at me from all sides during my journal club presentations. I had given my well-prepared talk. I had used good speech and volume, my body turned outward toward the audience, only referring to my slides occasionally to emphasize or explain a plot or conclusion. What more did they want from me? Now they had *questions*? How was I supposed to think of the answers to those questions in 0.0 seconds? How was that even possible? I started to dread the Q&A portion of the program.

I had a trifecta of issues that created fertile ground for that notorious impostor syndrome to take root. I was an African American woman in a field dominated by white men. I was an older returning student. And I was a classically trained actor. The impostor syndrome didn't just visit. It pitched a tent, had a cookout, started planning parties, and hosted ritualistic dances, howling at the Moon on my doorstep. The world wasn't big enough to hold my fear, of not being smart enough, of being found out to be the fraud that I feared I was—an actor in scientist's clothing (even if it was more fashionable clothing than the average scientist might wear).

Actually, feelings don't matter very much. Feelings come and go all the time. What mattered about feelings, about that fear of inadequacy and inaptitude, was what I did with them. This time around, I didn't hide.

I reached out for help from Eric Agol, a professor in my department. I shared with him, a white male and one of the most encouraging and supportive astronomers in the field when it came to inclusive practices, how frightened and inept I felt when it came to answering questions during and after talks. He told me to take some time after hearing the question, even a few seconds, to consider what was asked, before responding. If I had some idea of what to share, I could respond with that information. If I didn't, it was always better not to guess, but to say that I'd give it some thought and get back to them.

This approach sounded so easy; in practice, it wasn't. But gradually, it started to become second nature. The more familiar I became with my field, the easier it became to offer that knowledge and background as part of my responses to questions after talks. I learned other strategies for answering questions—even wild, off-topic ones—and still use those. Where once I felt victimized by questions from audience members, I began to feel empowered, even excited, by them. I could see that questions were how I knew that people were interested in my work. If you've ever given a talk and had *no* questions at the end, you know how defeating that can feel, leaving *you* with all the questions. *Did I talk too long? Is everyone just tired and ready to go home and eat lunch? Did they not understand a single word I said?*

I had support systems coming out of every nook and cranny, because this time I pursued them. Process group (aka "safe space") for graduate women of color on campus run by the mental health center? Sign me up. MS PHD'S mentorship program for minorities in Earth system science, created by the first African American woman to obtain a PhD in chemical oceanography from Texas A&M University, Dr. Ashanti Johnson? Check. I sat in a room with thirty other graduate students of color and for the first time in my life was not in the minority. If a white person had walked into the room, they would have been the minority! My chest expanded to take up my space in the room. I sat in session rooms at the twenty-thousand-attendee conference of the American Geophysical Union in San Francisco with NASA mission program director and Black woman Claudia Alexander on my right and a fellow Black Earth scientist from my mentorship program cohort on my left, and I was happy. I was not alone.

A mentor in that program, a female chemical engineer from Ghana named Dr. Akua Asa-Awuku, with smooth, deep-brown skin and a gentle and strong voice that soothed me immediately, told me to view my theater background as my superpower. Up to that point, I had assumed that I needed to sweep my unseemly foray into the humanities under the rug if I was going to be taken seriously as a scientist, and I'd acted accordingly. With a single comment, my mentor had turned my head around and pressed a button. The light bulb went on.

As the years went by, I read more papers—treatises written by others—about the research they'd cared about in some aspect of planetary habitability or climate or solar system evolution. And I would later write my own. When I was asked questions in a talk, sometimes I would have an answer for them. I often pulled a paper or two out of my mental back pocket and mentioned the possible implications as they pertained to their questions, based on this work. Sometimes I said I didn't know and left it at that. I didn't always handle every question the way that I would have liked to. It's like they say—there are three versions of every talk: there's the talk that you plan to give; there's the talk that you actually give; and there's the talk that you wish you'd given. Make no mistake, a talk—even a science talk—is a performance. And something goes wrong every single night of a performance. Even on Broadway.

This time was the beginning of getting to know and accept myself as someone who is a scientist—brown skin, makeup, nice clothes, refined speech, and all. I had never met anyone like me, so I had to become my own role model. But I was perfectly capable of doing that.

I remember being floored when I first read Carl Sagan's *Contact*. It is a story about a female scientist searching for life elsewhere in the universe. She doesn't believe in God. It isn't logical. Science is her religion. Theories can be tested and supported by evidence. There is no measurable method of proving the existence of a divine power.

By the end of the story, Ellie (along with an international crew, although in the movie version, it's just her) has traveled through interstellar space, using spacecraft designed by extraterrestrial life-forms that sent a message with blueprint specs across space and time to humans. She walks on the shores of a distant planet orbiting another star trillions of miles from the Earth and its Sun. She has a conversation with one of the ancient life-forms. She returns to Earth forever changed by the experience—and with absolutely zero evidence that she had ever left her chair on the launchpad.

With no physical, scientific evidence of her journeying to another world—no photos, no measurements, no other data—she is left with only her memories, and a firm belief in the legitimacy of her experience. Her experience cannot be proven. Yet it remains as real for her as the Moon in the sky. This is faith—a complete trust and confidence in the existence of something, whether it can be proven or not.

I decided that I was where I was supposed to be. If I was back in astronomy, it was because something wanted me there. And until I was given information to the contrary, I could assume that I was there to be a scientist. No human being was powerful enough to tell me otherwise.

So I'm a scientist who loves acting, film, memoirs, poetry, painting. I love Mary Oliver and how she writes of saving yourself when the whole world cries at you to mend *their* lives, and how she asks what it is we want to do with this one wild and precious life. It is never too late to be what you might have been. But you have to get on with it. Don't wait forever, because forever isn't that long; forever can be shorter than you could possibly imagine.

Snowballs and Life

I worked hard. I'd been away from astronomy and physics for over a decade, and I had forgotten a lot. Some things I had never really learned. So there I was, doing this "as if for the first time," as Salome often described it in method acting class. Everything you do onstage, from the moment you step into the scene, is as if for the first time. Like a child discovering the world.

I remember when I first learned about Snowball Earth. I was sitting in one of my core astrobiology classes. There were students from many different departments in the class—astronomy, like me, Earth and space sciences (or ESS), atmospheric sciences, biology, oceanography—all pursuing a dual-title PhD degree in astrobiology and their home department subject. There was another Black woman in the class besides me. She was in ESS, and she had been a pilot in the Navy; it had been a while since undergrad for her too. We passed knowing glances across the table during classes. We were often confused at the same time when the professor sped through background material.

The professor started talking about a couple of periods in Earth's history, approximately eight hundred and six hundred million years ago. This was before the explosion of multicellular life, before large animals like dinosaurs and mammoths. During this time in history, the earlier parts of the Neoproterozoic, life was small, microbial, and single-celled. Yet it was there and abundant.

Something strange likely occurred during this time. Ice sheets advanced all the way to the equator, freezing over the planet. To a distant astronomer, the Earth might have looked like a large snowball in space.

Why did this happen? What caused snow and ice to cover our planet from pole to pole? There are no definitive answers, only theories. The continents on our planet were arranged differently back then. A great deal of land was located in the tropics, where it rains frequently. As rain brought carbon dioxide (CO_2) out of the atmosphere, that CO_2 reacted with the minerals in rocks, becoming part of the land. This process is called "weathering," and today it acts to help stabilize our climate. When temperatures climb, precipitation increases, and CO_2 rains out of the atmosphere and gets locked up into rocks, lowering temperatures. The rocks and minerals get carried to the oceans, then get subducted as oceanic plates move underneath each other. Volcanoes then outgas CO_2 from those plates, and that CO_2 goes back up into the atmosphere, warming temperatures.

With a ton of land at the equator, CO_2 may have rained out of the atmosphere faster than it was able to be outgassed back up. It would have caused a sharp decrease in surface temperature, possibly plunging the planet into a global ice age. There it remained for millions of years, waiting for volcanoes to outgas enough CO_2 through the ice to melt it. Once the threshold was reached, geologic evidence shows that it may have taken no more than several thousand years for the ice to melt, returning the Earth to warmer, pre-snowball temperatures. Some think it took longer, closer to a million years. It's an area of active research.

Life survived Snowball Earth. Even photosynthetic life survived Snowball Earth, as there are markers in the rock record of biology that used sunlight to convert CO_2 and water into O_2 and sugars.

SNOWBALLS AND LIFE

How could this be? If the snowball episodes were truly global-scale glaciations, all of life would have been cut off from sunlight by a presumably thick ice sheet. Sunlight can only penetrate 100 meters at most into the ocean. Put an ice sheet on top of it, and that would make it pretty challenging for photosynthetic life underwater to get what it needs.

Perhaps the ice wasn't a kilometer thick, as many think. Perhaps only a few meters. But some models find that it's quite hard to maintain global ice cover that's only a meter thick without that ice thickening up. After all, bright ice will reflect sunlight, and that lowers temperatures further. As temperatures get colder, more ice will form. This is the definition of ice-albedo feedback, and it's one of the primary mechanisms regulating the Earth's climate. It works in the other direction too. If there's less ice on the planet's surface, more sunlight will be absorbed than reflected, and temperatures will rise, more ice will melt, the surface will absorb more sunlight, etc. This is what's known as a positive feedback loop. More creates more, and less creates less. The CO_2 cycle has the opposite effect, which is why it's a negative feedback mechanism. More CO_2 results in the CO_2 being drawn out of the atmosphere, lowering temperatures (more creates less), while lower temperatures lead to less rainfall to draw down CO_2, and volcanoes continue to outgas CO_2 into the atmosphere, which absorbs near-infrared sunlight, increasing temperatures (less creates more).

Perhaps it wasn't a global glaciation after all. Maybe there were pockets of open ocean—refugia—around tropical island chains, and photosynthetic life sequestered itself in these oases and soaked up some rays.

This mystery intrigued me. It posed an important question for the Earth: How could life survive a deep freeze? But I also began to wonder—if such an extreme climate state was possible on a planet that orbited its star at a reasonably moderate distance to maintain an otherwise temperate climate, couldn't it be possible on other planets orbiting at a range of distances around their stars?

LIFE ON OTHER PLANETS

I couldn't work twenty-four hours a day. There was Seattle to explore—a rainy, lush city where everybody had a box garden in the front or backyard, and a chicken coop. The weather was difficult at first, especially coming from Southern California. I was a Sun girl. The first things Steven and I bought were raincoats and waterproof boots. These were valuable investments, and they lasted the entire five years we were there. By the end of that time, I had grown to love the moist air, the overcast skies, how everything grew and settled itself into deep emerald tones in the rich Earth. It was like going to school in a rain forest. There was no smog, and when the Sun did make an appearance, the blue of those clear skies was spectacular. It shimmered.

Apart from their deaths, stars' lives carry a profound importance. The primary energy source for most planets is their parent star—the central mass in the system, from which all orbiting planets assembled following the collapse of the protostellar nebula. The habitable zone is the region around the star where a planet could be warm enough to have liquid water on its surface, assuming that its host star is its energy source. Too close to the star and the oceans boil; too far away and they freeze. In the habitable zone, temperatures might be not too hot, not too cold, but just right. Naturally, this region is often called the "Goldilocks Zone." It's where we look for planets that might have liquid water on their surfaces. We don't know if they do—only that they might.

But the traditional habitable zone is not the only place where life might be found. For planets that orbit frigidly far away from their stars, life might still exist at the bottom of an ice-covered ocean, kept liquid by the friction created by planets and moons tugging on each other, the way Jupiter's moon Europa nurses a global sea through flexing and stretching as it orbits its massive parent planet, Jupiter. Undersea life could be fueled by the chemical energy it receives from hydrothermal vents—chimneys created by the collision of oceanic plates, causing the seafloor to open up and release superheated water rich in dissolved minerals and nutrients. A delicious feast for these "chemosynthesizers." Do you see why the traditional habitable zone is only the beginning of the story of where we might want to look for life elsewhere in the universe? We can't put all our eggs in the habitable-zone basket, or we might miss out on the biggest discovery of a species' lifetime.

Neither could I put every single one of my eggs in the basket called "graduate school." If it didn't pan out again, I still wanted to have a life to go back to, with a husband and friends. My best friend from Exeter, Yvonne, had moved to Seattle years earlier to work for Microsoft. Her husband still worked for the company, and she was beginning to discover a love for visual art. We took art classes together and wandered the aisles of art supply stores looking for India inks and the perfect mixed-media notebooks to fill. We explored a New Age bookstore in Roosevelt and met to knit in a cozy nook outside an indoor shopping plaza in Wallingford. We devoured soft, warm croissants fresh from the oven with steaming hot chocolate to dunk them in at a pastry shop in Ballard, and we had herb cheese gougères with perfectly paired red wine in a small tavern in Wedgwood. Everywhere we went was full of people in sweaters and scarves keeping warm from the clammy chill. Before I left LA, someone told me that the cold in Washington went all the way through to your bones. They were right. Except in the summer, when the Sun came back. Then we toured the vineyards of the Columbia Valley for my birthday, and I got so tipsy I forgot about my fears. That felt so good that I'm surprised I didn't stay drunk for the rest of graduate school. Luckily, I wasn't wired that way.

Steven and I brought our love of bed-and-breakfasts to the Pacific Northwest and took every opportunity we had during the summers or on anniversary weekends to explore the surrounding communities—Bainbridge Island, Whidbey Island, the San Juan Islands, the Oregon coast. I gobbled up Dungeness crab, warm and much sweeter than the crab I'd tasted in salads back in California. We walked the narrow cobblestone streets of Pike Place Market, sipped hot bowls of clam chowder underneath waves of passing seagulls on the docks. We watched fishmongers as they threw fish to each other behind the counter for the tourists' cameras. We discovered the large, rectangular packages that were the crab rangoons at Mee Sum Pastry, full of cream cheese and fresh crab and wrapped in only the thinnest wonton shells. Steven fell deeply in love with the coffee, as he was bound to. I found my elixir in genmaicha tea from the Perennial Tea Room and bought it by the bag, opening the top each morning to inhale the nutty smell of toasted rice mixed with fragrant green leaves.

I was eager to fit in, and I followed my classmates out on Fridays for "whine time," which was usually followed by dinner at some hip restaurant. This was a foodie town, where everything—from paneer to seafood to pizza to wine paired with local cheeses—was done well. One Friday evening before Halloween we were all out at a Middle Eastern restaurant when my cell phone rang. It was Steven.

"I thought we were going to carve pumpkins tonight."

Suddenly it hit me. *I'm married. I'm not twenty-four. I need to check in.*

While I couldn't have been busier during those first two years, Steve couldn't have been less busy. He was on unemployment for the first year and a half that we were in Seattle. He had left his great-paying job at the Pantages Theatre box office in LA, where he could go on auditions anytime (a truly precious thing for an actor), to sit at home and look for work and wait for me to finish problem sets so that we could watch movies and live like a regular married couple. Except we weren't a regular married couple. Wives uprooted themselves for their husbands' jobs all the time. It was expected. Yet Steven's doing it for me . . . I felt enormous guilt over it all. When he got a job in the Seattle box office scene and started to find a theater community and a buddy to sample coffee with, I rested easier.

☾

I wrote a poem called "Nefertiti with a Calculator" that was published in the campus literary arts journal. A line in the poem read, *What do you do when your own mind is your biggest racist?* A reporter once asked me what I meant when I wrote that line. I responded with some intellectual answer, trying to make sense of it several years later. It wasn't the truth.

I wrote the poem the spring before I took the qualifying exam—that six-hour written exam I mentioned earlier that covers thirteen courses and two years' worth of fundamental astronomy and astrophysics knowledge. It was an exam that, as a second-year PhD student, I was supposed to pass. The last Black woman in the department who'd taken this exam had failed, and then failed again, and was asked to leave the program. And there I was, poised either to repeat history or to make it.

Someone once said that a poem is a moment moving through you. That's what that line was for me. I don't remember what I meant when I wrote it. I was probably thinking of my experience at UW–Madison, and of how high the stakes felt here. What I know about that time is that I was scared, I felt alone, and I desperately, with every shallow breath I expelled that year from my narrow throat passage, didn't want to fail.

Before taking the qual, I'd also gone back to Taos. It was the first retreat I'd done with Natalie since leaving Los Angeles for UW. It had been years. I was aching for silence, so I wrote to her again. She offered me a partial scholarship, and I flew from Seattle to New Mexico.

So much of it was the same—the sitting, the slow walking, familiar students who'd kept coming to workshops during the years I'd been doing problem sets. I slid back into the structure and let it fill me. While I'd been gone, I'd read more of Natalie's books, listened to her on tape. In one of her talks, she'd mentioned how important it was to choose something and go deeply with it. I, of all people, had finally chosen. But I was scared that I wouldn't be able to succeed at the thing I had chosen. I hadn't proven myself yet. The qualifying exam loomed large ahead of me.

In class Natalie gives students the opportunity to write down questions on slips of paper and leave them in a pitcher for her to read and respond to later, often privately. I wrote several paragraphs in tiny script so that it would all fit onto one side of a piece of paper. I told her about the Everest of mountains that it felt like I was climbing just being back in astronomy, the weight that I carried for fear of failing as a Black woman in this field. I laid it all out. I folded up the paper, placed it at the mouth of the pitcher, and let it go.

The next morning, I walked into the zendo and over to my meditation cushion to get ready for class to begin. A folded slip of paper rested on top of the cushion. I picked it up, sat down, and unfolded it. On the paper was a single line Natalie had written in response to my diatribe.

You're on the horse. Now fucking ride it.

It was exactly what I needed.

The Qual

Toward the end of the movie *Legally Blonde*, Reese Witherspoon's character is about to leave Harvard Law School to return to LA. She's completely lost faith in herself and her potential, and is ready to throw in the towel and head back to what's familiar and known. She says to Luke Wilson's character, "No more trying to be something that I'm . . . I'm just not." He replies, "What if you're trying to be something you are?"

It was June, and I was walking to school to take my qualifying exam at the end of my second year. It was to be the second time I took this exam. The department administers it for the first time at the end of a graduate student's first year. Since the exam covers two years of courses, students aren't expected to pass it the first time. A precious few do, but in general it's meant to give them a chance to practice sitting still for six hours, with a one-hour break for lunch, during which they aren't allowed to run to their offices to pull out notes or get on the internet to check answers. Mostly the department wants to scare the shit out of them to ensure that they'll study their asses off the next time around.

I'm a terrible crammer, so I knew I needed to start studying early and small. First, one hour a day. I started seriously studying, at least two hours a day, after spring break. The qual was in June. I was terrified of failing. If I failed I would get one more chance to pass the following year. Waiting an entire year to take this fucking test sounded like the slowest kind of medieval torture—like the one in *Game of Thrones* where rats in a bucket eat their way through people's abdomens into the intestines to escape fire at their heels. I had to pass this year, for my own sanity.

This time I had shown up to those group study sessions, instead of politely declining and flailing on my own as I had done in Madison. I was leaving nothing to chance, or to my natural inclination to isolate. I had gone through each of the thirteen courses in painstakingly organized detail—personal lecture notes, old problem sets, tests. I filled two index card boxes with vocabulary terms and equations—the equation of radiative transfer, governing how light propagates through a medium; the Eddington limit for a star, which determines the maximum luminosity it can have while still remaining in hydrostatic equilibrium, where the radiation pressure pushing out is balanced by the gravitational force pulling in; the white dwarf cooling rate. And so many more.

I had followed another tip from my former colleague Kartik at Caltech. After each lecture, back in my office, or on the bus ride home, I turned to a fresh page in my spiral class notebook. I'd make three columns. The left column I titled "Keyword(s)." At the top of the middle column I wrote, "In my own words." Then at the top of the right column, "What I don't understand." I distilled the day's lecture down into no more than seven key points. I wrote each one in the left column (for example, "maximum CO_2 greenhouse"), then described them ("the outer edge of a host star's habitable zone, beyond which any further increase in CO_2 is no longer sufficient to keep temperatures above the freezing point of liquid water"), along with any relevant equations, in the middle column. Whatever I was unclear about regarding that topic went in the right column ("What exactly happens when this atmospheric CO_2 threshold is reached? Why,

THE QUAL

at a certain distance away from its star, doesn't any further CO_2 volcanically outgassed into a planet's atmosphere increase temperatures?"). That right column was what I brought to the professor's office hours. That was how the items in that right column disappeared. That was how I got to understand that once enough CO_2 molecules get into a planet's atmosphere, those molecules start to scatter starlight, and that scattering of light away from the surface (cooling the planet) eventually begins to dominate any warming effects caused by CO_2 absorbing that light.

Now all of these three-column pages, distilling the chicken scratch notes I'd taken in class into tight, clear lecture summaries, had become built-in study guides for the qual. And making them had been a way of studying as I went. The courses I had done the best in were the ones in which I had followed this practice religiously.

Even more critical, I had finally done what I had never understood the significance of doing at MIT or Madison—drilling problems. It didn't matter how many equations I memorized, how many problems and their solutions I went through and understood *theoretically* in my head. If I didn't put pencil to paper and actually do the problems myself, I would end up staring at an exam with no idea of how to begin. Then I would start throwing equations randomly at a problem, hoping I could plug numbers in and get something close to the right answer, in the right units. There were only so many types of physics and astronomy problems that could be created to conceptualize a principle. There were only so many ways you could draw a three-body diagram, and only so many variables the problem could ask you to solve for, using that diagram. There were only a few ways to write out the equation of radiative transfer, and about as many ways to solve it. There were only a few main density regimes in the interstellar medium. The boundary between a planet and a star was clear and clean. The boundary between the physics used for the interior of a star and that governing its outer layers was less clean, but there were guiding principles with underlying assumptions to be adopted in each regime, and these were always clearly stated in a problem. If I did enough of these problems, in enough different ways, I would see the pattern, the repetition, the single small element that had changed between one problem and the next. *Oh yeah, this is that conservation of mass problem again. I've seen this before . . .* They would become predictable, even easy. I had never done enough problems before to get to that point.

I had a binder filled with qualifying exams from prior years. We all took these practice exams together, then went over the answers as a group, writing the solutions on chalkboards in the same conference room in which we'd taken each of those thirteen courses, asked and answered questions from our professors and each other, felt our eyes glaze over and refocus in lecture. Now the cord had been cut and we were swimming out on our own.

On exam day I sat in a tight desk—one where the table is attached to the chair—with a pile of paper spread out before me. Two hours into the six-hour exam, my hand was furiously scratching across the paper, pulling out and depositing onto each page the contents of my brain.

During break, the faculty organized lunch to be brought in for us. We all sat in a room together and tried to talk about something else, anything but the problem we were in the middle of solving on page five, the one doing a merry-go-round in our heads, the one we couldn't wait to get back to our own desks to finish before the knowledge left our brains forever. Someone put an international televised soccer match on the big projector screen in the room and turned up the speakers. We quietly munched on our sandwiches and chips amid the cheers and vuvuzelas, our best attempts to drown out thoughts.

The last problem finished itself through the equations autopiloted by my fingers. The last of the hourglass sand grains ran out. It was over. I put down my pencil and stretched my fingers wide. I exhaled. All I could think of as I brought my papers into some order and walked them to the front of the room was what I had left undone.

I walked out with my classmates, talking jovially about it being over, how we couldn't wait to drink now. "I wonder what the third-years have planned for us," Libby said. The third-years, all of them having presumably passed their qual, had the honor and responsibility of planning the "qual party" for the first- and second-years who had taken the exam that year. We would partake of all the food and alcohol that found its way into the reading room—the place where we took classes, listened to talks, and, it turns out, would get plastered after enduring six-hour exams. I walked to the reading room slowly, deliberately, as if I wasn't quite on the ground, but somehow hovering an inch above it in suspended animation, while the rest of the world continued on.

I didn't get plastered. I drank a few plastic cups of wine and left my younger twenty-something classmates to continue their debauchery into the night, and probably into the next morning. I got on the bus for the short ride home. My father called. He was about to leave on tour with my mother, a reunion of the Pyramids; they were scheduled to leave for Europe in a matter of weeks. "I can't get ahold of your mother. She's let me know she's not doing the tour." I listened with half of my mind. I had just talked to my mother, so I knew she was fine. I knew my parents had had artistic differences before, and that this was one of those pre-tour dramas I had no business getting involved in. I told my dad I hoped they could work it out. But I had no more to give. I had just taken an exam that would decide my fate for the next year of my life. Either I would get to spend it starting my dissertation, or, if I didn't pass, I would spend the entire year with the knowledge that I was on probation, and everyone in the department would know so too. I could keep doing research, but I would have to keep studying those courses I'd already taken to keep the material fresh in my mind. I wouldn't have earned the right to let it all go and dive in fully as a researcher. I would get one more chance to pass the qual, but I would have to wait until the following June to take the exam for the third time.

I thought of the stories I'd heard in the department about Kate, the last Black woman to be a grad student in the astronomy department. After finding out that her second attempt to pass the qual was unsuccessful, she had demanded to retake the exam immediately rather than wait the requisite full academic year. She was appeased, but it was not in her best interest—a prime example that sometimes the things we want in the moment are simply not good for us, no matter how much we want them. After failing the exam for the third and final time, she was invited to leave with a master's degree. Another Black woman in recent department history had needed a third attempt to pass the qual. Her final attempt had, fortunately, been successful.

I felt heavier and heavier. I feared confirming stereotypes about my race and gender if I failed. Hearing the harrowing stories of the Black women who'd gone before me and struggled desperately with this monstrous qualifying exam—one finally succumbing to its power—did not help my fear. It turned it into a near certainty.

When I got home after taking the exam, I went right to the refrigerator, planning to only peruse. There was a cake Steven had made just for me—vanilla with vanilla frosting ("Yeah, you know it. You love vanilla. And I love chocolate," he always joked)—and written on the top, "Congratulations" in piped frosting. My heart felt warm. Still queasy and unsteady in my chest, but snug as a bug in a rug.

The morning after the exams felt like a funeral march. I walked down the hallway slowly, deliberately. The walls constricted, closing in and then expanding, like in a carnival fun house. In the hallway as I approached my office, I could barely make out a group of my classmates hovering in the doorway. They were smiling, their bodies relaxed and animated. They looked like they were in those dizzy moments after getting off a roller coaster. Every cell in their bodies seemed to sizzle with electric energy and gratitude to be alive and back on solid ground.

I tried to look relaxed too. But my face felt tight and contorted. When I got to them, I saw on the right periphery that Scott, our Radiative Processes instructor and the head of our graduate program, was walking toward us. He looked like he had been watching for us, circling the runway, waiting for a signal to touch down. I learned later that he'd been down this hallway many times already that morning to tell each student in turn the results of their exam.

When he got to us, I almost wanted to apologize right then and there, before he spoke. I wanted to tell him that I really did know all the answers; I just ran out of time. I wanted to tell him that I'd never been a very good test taker, but my instructors knew I did good work on problem sets. *I am thorough*, I wanted to tell him. I wanted him to know that I even got an A+ in Extragalactic Astronomy last quarter. Maybe then the news he was about to give me wouldn't hit him so hard, wouldn't make him feel too disappointed in me. Maybe he would even go back and ask the faculty to regrade my test. Maybe they missed a page.

I held my breath as he approached and gave him a nervous smile that said it all—*I'm scared of what you're about to tell me, and I'm not sure how I'll react if you say what I'm most afraid you are about to say to me.* I waited for him to speak.

Scott was a thoughtful, kind, and deliberate man and speaker. When he had something particularly important that he wanted to say, he was emphatic, and his first words were often slightly stuttered coming out. He looked directly at me and said, "C-C-Congratulations. You have passed your qualifying exam."

"Are you serious?" I asked him. For a moment I thought I'd imagined it. The other classmates had received their news during one of Scott's earlier flybys that morning. I felt hands on my arms, my shoulders. There was a cacophony of congratulations in stereo. My eyes met Scott's again, still in shock, amazement, and wonder, not yet crossed over to elation. That would come later, after the waves of disbelief and suspicion that perhaps they'd mixed up my qualifying exam with someone else's had dissipated.

As much as a year later, I was still waiting for the chair of the department to come to my office to tell me that I would have to take the qual again the following year. I fantasized about the level of keening I would do at her feet as I begged and pleaded for my life. It would rival the greatest productions of Greek tragedies that had ever been mounted, in actual Greece, during ancient Greek times.

But no chair ever came to my door with that news.

It's strange how things can shift. It can happen fast, like an iceberg in the Arctic seems to all of a sudden break off and plummet into cold waters. Or it can happen slowly, like the shifting and moving of continental plates. One day the whole world is in a different configuration, and you never saw things move.

I remember patchy feelings of old worry and fear. I moved on anyway, in the next stages of things a graduate student in our department had to do. I proposed a dissertation topic and wrote a proposal. I selected a dissertation committee. These were professors at my institution—along with one outside member who could be more impartial, for fairness—who agreed to read my proposal. Assuming all went well, eventually they would read the dissertation itself—a book-length manuscript I would write on my topic of choice. This manuscript would be stitched together from the papers I would write and publish in peer-reviewed journals over the course of three years.

THE QUAL

The committee would sit in and preside over the last two big exams I would ever take in my life—my general exam and my final exam (aka my dissertation defense). I would walk through these rites of passage, one by one, with the help of those who had come before me in my department. Ours was a department steeped in tradition. For the general exam, I was given the topic one week in advance. It was a topic that was orthogonal to my dissertation topic—a topic that the committee (which was chaired by my advisor), given my dissertation proposal, thought I should know something about. Since I was all about planets that could get cold, the committee gave me the opposite end of things: how planets can get hot—so hot that they lose all of their oceans as the result of a "Runaway Greenhouse" state, like that on the planet Venus. Over that week I read every paper I could get my hands on that covered the Runaway Greenhouse effect, and put together a forty-five-minute presentation. For my exam, this presentation would be given to a public audience, followed by the opportunity for anyone in the audience to ask questions. Then everybody else would leave except for the dissertation committee, who remained behind to ask additional questions for an hour in a "closed session." The committee would then kick me out of the room while they deliberated about whether I passed. If I passed my general exam, my graduate student status would be elevated to "doctoral candidate." I would then be free to continue with my research and with the writing of my dissertation. The next time all of us would be together again would be on the day and in the designated room of my final exam—my dissertation defense.

There was a tradition of giving a practice talk in front of classmates and other grad students in the department two to three days before giving the general exam talk, and about two years later, the defense talk. This was a chaotic time, when I had barely strung together enough to halfway resemble a coherent, lucid, and authoritative presentation. But my classmates were unflinching, and lovingly, helpfully critical. They pulled no punches. None had been pulled for them. Or they knew they would one day be exactly where I was, and they wanted the truth to be told to them on that day. Hour-and-a-half diatribes transformed into tight, forty-five-minute, story-driven journeys, thanks to the brave editing feedback of these practice talk sessions and the good sense of those of us who took our turns in the hot seat to implement said feedback over the final few days before we did this for real.

I walked to school every day during this time. On my right, across the deep chasm of the reservoir, and past several Seattle boroughs—Green Lake, Wallingford, Ballard, Shoreline down Aurora—I could see the outline of the Olympic Mountains. The air was clear, the sky a sparkling bright blue, with no hazy, milky orange film. Not like LA. The mountains stood tall and etched the sky, their dark shapes forming some alien body on its side, far off in the big world. They seemed like a distant landscape apart from my life. I heard only the steady stream of passing cars on my left, racing down 15th Avenue NE toward the University District, where I was headed.

I recited poems I'd memorized day after day on these walks. They were poems that inspired me to get up each morning and try again, every day, to be better at this than I was the day before. They were poems that reminded me that it was only me who could set a line in the sand that would mark how far I was capable of progressing in this field. And they were poems that told me that I was the only one who could wipe that line away with my foot or the back of my sleeve, because it never should have been there in the first place. They were the poems of saving myself by Mary Oliver, of the clear bell of *I can* by Denise Levertov, of the daily joy that just being alive can hold by Jimmy Santiago Baca, and of being the captain of my own soul by William Ernest Henley.

I have never met their authors. But they are the poems that are my kindred, my breath, my salvation. They reminded me that I was alive and I was whole just as I was. And I wanted things, and that was fine, so I should just go get them and stop whining about not being good enough already. I was in charge of those efforts, so I needed to get to work.

And so I did.

Ice

For my dissertation, I knew that I wanted to study something about planets around other stars, though I didn't know what, exactly. I needed to come up with a scientific question to try to answer. While I had been gone from astronomy, the field had exploded. In 1995 we knew of one planet orbiting a main-sequence star outside of our solar system. Now, in 2011, there were over a thousand.

And there were at least that many questions about each of the planets that had been discovered. Were they rocky, like the Earth? Or gaseous, like Jupiter? Or were they somewhere in between? Why had they formed as they did, in the arrangement they did? What were their atmospheres like? Their surfaces? Were they capable of supporting life? What kind of life? Did any of them host life now?

I was curious about all these questions, and addressing any one of them would require multiple dissertations, if not lifetimes. I needed to find something specific and bounded to study, and it had to be something that didn't just make me curious. I needed to be personally invested. I had gotten to the point, upon passing the qual, where I had started to believe that I could do this—this science thing. If I worked hard enough at it,

I saw results. It was math. A (my mature, thirty-four-year-old self) + B (working my ass off) = C (getting the grades to move forward). I knew it would be the same thing with research.

But all of that was external. I had been basing my ability to succeed on external metrics for success—grades in courses, grades on exams (including the big, scary qual). I'd been thinking like this for years. Research, too, had been and would continue to be that way, my success based on my advisor's assessment of my progress—how well I learned to use a given research technique or programming language, how often I produced results and shared them in group meetings and wrote them up for publication, how well I communicated my results to the broader astronomy community at conferences.

But this dissertation was something that I would be doing, day in and day out, for the next three-plus years of my life. I knew that I would need to find more to sustain my enthusiasm for the topic than the external markers of success that were inherent to my continued progress and assessment. My advisor, my peers, the larger exoplanet community—none of these were going to be at my bedside at 7 a.m. every day of those next three-plus years, whispering in my ear to wake up, eat breakfast, take a shower, and get to work. I was going to have to do that all by myself. It needed to be personal.

I walked into journal club one afternoon and a senior grad student was presenting a paper. The air went very still, like it did all those years ago in the seventh grade when I saw *Space Camp*. The paper was about how ice on a planet could behave differently depending on the type of light it received from its star. In my mind ice was a very bright, white thing that made other things—like the water in the glass next to me as I wrote this—cold. Because it was bright, ice reflected lots of light away from it, cooling the surface even more. It was why "Snowball Earth" had happened those six hundred million years ago, and two hundred million years before that. Ice-albedo feedback, a positive feedback loop, is powerful. It would amplify, rather than dampen, the original thing that started the whole process—a little bit of ice on the ground. And all those years ago, it most likely turned the Earth into one big snowball in space. Remember?

But according to this paper, ice could be very dark and absorbing if the light that hit it was in the infrared part of the spectrum rather than the visible, as most of the light is that we receive from the Sun. Ice wasn't always a bright white thing after all. What ice on a planet was, what it did, depended on the kind of light it received from the planet's host star. And that type of light depended on the particular type of star itself. Infrared light was what stars much smaller than the Sun emitted most. And this longer, redder wavelength light was the kind of light that ice loved to absorb.

I thought about this "bipolar" nature of ice. It was like me. There I was, smack-dab in the middle of an astronomy PhD program, hoping like hell that all this work wasn't going to be for nothing, and yet a ring around the edge of my life was alight with the world that I had most recently come from—a world of stories and emotions and characters and feeling. I had returned to astronomy because I wanted to be there. But I had changed since the last time I was in astronomy. I wanted to bring what I'd learned to my study of the universe, and had been silently tapping on a door for a way in. This was it. Even in its extraordinarily technical justification, it was personal. A hypothetical planet orbiting a Sun-like star at a certain distance might be a frozen wasteland, its ice snowy white. But put that same planet around another star, and its dark-gray ice might make it warm enough to harbor oceans teeming with life. I loved this apparent contradiction, how ice can completely challenge the way we think of it, how it can do one thing and then turn around and do the complete opposite, simply because of how it's made. The molecules of its phase of water vibrate against each other in a certain fashion that demands that ice behave in this way.

This paper that the senior grad student was talking about proposed that if you put a planet with ice on its surface around a small, cool, red star, and you put another planet with ice around a large, hotter, bluer star, different things would happen on each planet. If both planets were orbiting their stars at the distance where they received the same total amount of light, the ice on the planet orbiting the redder star would absorb more energy from that cooler red star because most of its light is infrared light. The ice on the planet orbiting the hotter, bluer star would reflect more energy away because most of the bluer star's light is in the visible and UV, which ice likes to reflect. In other words, planets orbiting cooler stars might actually be warmer than planets orbiting hotter stars.

I was captivated.

I had no qualms about emailing these authors I didn't know. I had been an actor. Actors have to introduce themselves to strangers at parties—and in cold, sterile audition rooms, in front of deadpan, bored, doubtful faces almost daring them to be the one to give them a reason to go home early. I knew how to market myself. So that's just what I did.

This paper had not included any computer modeling to support the authors' argument, and I wondered if they had plans to do any other work on the subject. I emailed the paper's first author, Manoj Joshi, asking him if he had planned to do more work on the effect of host star spectrum on planetary ice-albedo feedback, and if so, could I be involved. He responded right away. He told me that he had no plans to do future work on the topic and that I was free to take the idea and run with it. All that he asked was that he be a coauthor on any future papers I wrote on the subject.

So began my dissertation.

I am standing on a stage. It isn't 1998, when I left the field the first time to become an actor. It is 2012, and I'm a second-time astronomy graduate student. These are the finals of the U.S. version of FameLab, an international science communication competition where people from all over the world explain complex scientific concepts in understandable and entertaining ways, in under three minutes. I'd won the regional competition in Denver earlier in the year. Now I'm in Atlanta, Georgia, where the Astrobiology Science Conference (aka AbSciCon) is being held. The FameLab finals are happening the day before AbSciCon starts.

In the days before the competition, the other finalists and I arrive at the hotel site for a workshop on communicating science. We all know each other's two-minute pieces by heart. We'd been workshopping them all weekend. Since the time limit is absolute, everyone has memorized their pieces. In the workshops we talk about projecting our voices (even though we will all be miked), making eye contact, how to hold our bodies, how to stand. Some finalists will use props to convey their message— like a moon rock or a circuit or a seashell. I've decided that I don't want anything in my hands. My hands are my tools, along with my head, face, arms, chest, hips, and legs.

My piece is called "Bipolar Ice." It is all about the dual nature of water ice and its implications for the climates of planets orbiting different types of stars. How ironic that I should be obsessed with icy planetary surfaces for my research. I get cold when it's 65 degrees outside. I regularly bake Steven at night blasting the heat. Still, I dream of the majestic glaciers of Alaska, of hunting for meteorites on the barren white plains of Antarctica. I know in my bones that I would be ready to go home on day two, from both the cold and the lack of indoor plumbing. Yet it sounds so cool.

ICE

I can't just tell the audience—which fills two conjoined ballrooms and contains my husband, father, aunts, uncles, and cousins, most of whom live here in Atlanta, along with my advisor and the director of the NASA Astrobiology Institute, which is sponsoring this whole competition—that ice has this property and that's why it's fun to study. I need to show them. That old adage still stands: "Show, don't tell." Whenever you have an opportunity to show someone something rather than tell them, take it.

So the ice becomes a character. The ice on a planet orbiting a star like the Sun talks like a valley girl, saying, "Thanks, but like, no thanks!" to the Sun's light, sending it packing, back out into space and away from the planet's surface.

I hear laughter and I know I've got them. I've made this something they can understand. I haven't hit them with technical jargon, or, God forbid, an equation. I've personified my science focus, given the ice feelings, needs, wants, desires; I've made it something any person—science literate or not—can relate to.

The ice on the planet orbiting the cooler, redder star is a lover, the smoothest, most sensual, and sexiest woman on Earth. She can romance any man or woman, married or not. To the red star's light, she says, "Come on in, baby. I need you, I want you, I've got to have you. Don't worry, baby! I'm not going to reject you and reflect you. I'm going to absorb you. That's how much I love you . . ."

Whistles, laughter, clapping. My body tingles with that familiar feeling—pride. *God, how I love to perform.* And here, this moment defines the singular event of both sides of my brain being active at the same time. Both parts of me represented. It is as if I was made for this moment. There would be others, but this is the first.

Once I had them, I could give them the science implications and know they would hear me. There might not be much in the way of ice on planets orbiting cooler stars. There's the irony, again. That longer-wavelength light they emit makes a difference. The more of that light the planet absorbs, the warmer the planet gets.

Nichelle Nichols emcees the competition. After she introduces me, saying my name perfectly (I am always listening for this), I hear her say, "You go, girl." On *Star Trek* she played Nyota Uhura, one of the first African American women to be in a major television series. She was on the bridge in that show, every single episode, translating alien languages heard over special frequencies on an interstellar spaceship hundreds of years in the future. She tells a story of how she thought of leaving the show at one point. Then she met a young Dr. Martin Luther King Jr. at a civil rights rally and he urged her to stay, telling her how important it was for Black and white people everywhere to see her on TV, in that role. So she stayed.

In the hotel lobby after the finals, Steven and I run into Ms. Nichols on our way to celebrate with my family. Steven thanks her for being who she was all those years ago on *Star Trek*. Without her, without that first-ever interracial kiss with William Shatner, which broke open the world and everyone's ideas about what was and was not okay to see on television, we—he and I—might not be here, holding hands in public, being married. He gets teary as he says this to her. She hugs him.

My family sitting in the back clapped and cheered the loudest in the audience. I sat with the other contestants as we waited to hear the results. The audience got to vote for their choice, and the three judges would determine the overall winner after a short public Q&A with each contestant onstage after their piece. I don't remember what question they asked me or how I responded. That's testimony to how far I'd come. It must not have been that big a deal.

I won the Audience Choice award, but I was not the overall winner. That honor went to a man who'd also been funny and talked about life and probability. My father and extended family told me afterward I should have swept both awards; I had random strangers come up to me over the next week during the conference to tell me the same thing. It was like I was the Al Gore of FameLab. I was okay with that.

Later in the week my poster on the effect of host star spectrum on the latitude of the ice line on orbiting planets won first prize in the conference poster competition. In the picture taken of me standing in front of my blue-ribbon poster with NASA Astrobiology Institute director Dr. Carl Pilcher on my left and Dr. Mary Voytek, NASA senior scientist for astrobiology on my right, I am beaming. I'm also wearing mascara.

The Weather on Other Worlds

In Southern California, I usually don't have to worry about the need to wear a parka. But as a child in western Massachusetts, we had glorious things called "snow days." In the dim early light I would wake to the choppy sound of my grandmother's radio on her bedside table in the room next to mine. I'd scramble out of bed and sit on the floor at her feet. Together we would watch the black square box, willing the announcer with our eyes to call out the name of my school. When I would finally hear Amherst Regional Junior High, it was like a Hostess Twinkie and a cupcake and the last day of classes before summer recess all rolled into one. Magic.

I wanted to find a way to tell what the weather could be like on this long list of planets that had been discovered. I didn't care that much about the big, Jupiter-sized ones. Those were gas giants, and no one could walk around on those. I mostly cared about the planets that were close to Earth's size. Those were the ones that made my insides leap around when I heard their names. Not many people were trying to figure out what the climates of exoplanets might be like. Most exoplanet astronomers were focusing on finding the planets to begin with, which was a difficult enough job, considering how far away and how small and dim they were compared to their host stars. They were easily swallowed up in the bright glare and dominating gravitational force of their parents. Most children are, at some point. But these children were located trillions of miles away and were about as bright as a fruit fly crawling around on a headlight—trillions of miles away. Exoplanet observational astronomers definitely had their hands full.

I'm thinking of the time I met Ray Pierrehumbert—then a professor of geophysical sciences at the University of Chicago—for coffee. It was at a big annual conference of the American Geophysical Union. He was the first person I knew of to take climate models used to predict climate on the Earth and apply them to exoplanets. He had discovered new ways that planets could have liquid water on their surfaces—exotic, alien ways, like only having an open circle of ocean on one part of a planet, with the rest of the planet frozen around it. The planet could look like an eyeball. He called this planet an "Eyeball Earth." He and Dorian Abbot at the University of Chicago wrote a paper about how continental dust collecting on snow and ice during Snowball Earth may have relaxed the amount of CO_2 required to build up in Earth's atmosphere for it to melt out of global ice cover. When Manoj Joshi had told me I could run with his suppositional idea of a suppressed ice-albedo feedback on M-dwarf planets, I'd still needed a plan. I wanted to understand how all of this fit into the larger field. What had been done already, and where were the gaps? So I emailed Ray. I'd met him briefly at the Aspen Center for Physics during the first meeting of Exoclimes—a small, discipline-specific conference for people working on exoplanet climates. During a public talk for the Aspen community he'd shared one of his favorite science fiction stories, "A Pail of Air" by Fritz Leiber. It would become one of mine.

In "A Pail of Air," the Earth has lost its star. It wanders aimlessly in the blackness of space, cold and rogue. A rogue planet. They do exist. Not every planet has a parent star whose light it can bask in, staying warm and clement for liquid water, for life. So in this mythology of Fritz Leiber's, the Earth without its Sun has gotten so cold that not only the water freezes, but the air too. The very atmosphere. Every kind of gas that comprises the atmosphere has condensed onto the surface of the planet. And the miracle of this science fiction story is that it is grounded in science fact. The order in which the gases froze out and layered themselves one on top of the other matches science. First the water froze out onto the surface. Then the carbon dioxide. Then the nitrogen. Then the oxygen.

But Fritz doesn't just tell you about a lonely planet in the black, cold and dead. We learn that this happened suddenly, through a family's eyes. We learn how they think they're the only ones still left alive, how they survive deep underground, where it's warmest. How their cave is insulated with blankets, and every day one of them goes up to the surface to fetch a pail of frozen oxygen to bring back home. The chunks of oxygen ice go over the fire, where they warm it into a liquid and then to a gas, so the family can keep breathing, until help comes . . . someday.

I sit across the table from Ray, watching the napkin. His large fingers on his left hand hold the napkin, while the silver pen in his right hand draws lines on its surface. He draws a graph, of the albedo—the reflectivity—of water ice at different wavelengths, alongside that of ocean water. As he talks to me his voice is deep, slightly gruff, but warm, repeating the beginnings of sentences before going on, because he is excited. I'd read some of the existing literature, and he gives me more names, stringing threads through the bullet points of new ideas that have emerged through the years explaining the advancing and retreating of ice on the Earth and other planets. His hair is graying and his face whiskered. I watch his mustache and beard ruffle in his breath. I'm hanging on to it, while my belly flutters in its pit at the space I'm in, sitting here with him, learning. It is all tying back into itself, forming a firm knot of an idea from the disparate threads of stars, ice, infrared light, Snowball Earth, and planetary systems. I see it there on the napkin. It is a story, like all things, like the pull of my entire life.

My advisor, Vikki Meadows, knew of a woman in atmospheric sciences who used climate models to predict the patterns of arctic sea ice on the Earth. She set up a meeting for the two of us. It was like being set up on a blind date. Would I like her? Would she like me enough to show me what she did? Would she want to work with me, using her models in ways she probably had never thought about before—not to study the Earth, but other possible Earths in other solar systems? How would we even go about doing that?

Turns out, like all things, the answer was to start small. The 3D global climate models, or GCMs, that Cecilia "CC" Bitz used to study the behavior of arctic sea ice over time had been used for decades to predict climate and weather patterns on the Earth, and to predict the worsening impacts of carbon dioxide–induced climate change caused by humans into the 2100s. They required inputs such as atmospheric composition, the orbital elements of the planet—how tilted the imaginary axis that extends through the north and south poles is (a planet's obliquity), how oval-shaped the planet's orbit is (its eccentricity), the spectrum of the host star. They could get complex quickly, including the options to incorporate soil cycles, vegetation cycles, and ocean circulation into climate simulations. Historically, all of these inputs had always been those that described the Earth around the Sun.

The surface albedos for ice that were input into the GCM also assumed that the host star was the Sun. We now knew that, just as the host star needed to change in the model, so did the albedo of water ice on the planet, because that albedo depended on the type of light emitted by its host star. Glaciologists like Stephen Warren over in UW's Department of Atmospheric Sciences had known this for years. He traveled to Antarctica regularly to set up special cameras on the barren ice sheets and measure the albedos of ice. He'd published albedo measurements for every kind of ice, from snow (the finest-grained form of ice), to striking blue marine ice, formed from freezing seawater and containing so few cracks that very little sunlight was scattered away as it passed through the ice. While the redder light was absorbed along the journey, it was blue light that made it all the way to the bottom of the ice, allowing it to be reflected back up to our eyes so that we saw it as blue ice. I met with Steve and watched him pull paper after paper of his out of a file cabinet in his crammed office and hand them over for me to read. He pulled out a yellow lined tablet, drew x and y axes, and wrote out an equation for wavelength-dependent albedo. It wasn't enough to know that water ice could behave differently toward different types of light; I needed to understand why. He showed me the math. He said I could use his albedo measurements in the GCM. He didn't want to be a coauthor on my paper. He had enough on his plate. But he would be a resource.

Climate models could be 1D, where the only dimension was Earth's latitude, and the atmosphere was essentially collapsed down into a pancake. Since the inputs were tuned to the Earth, 1D energy balance climate models (often called "EBMs") could simulate Earth's climate within minutes.

We knew what the Earth's atmosphere was composed of—mostly nitrogen, with oxygen, carbon dioxide, and some trace gases—and we knew what our surfaces were like. We had satellites up in space orbiting the planet and constantly taking data. We knew the Earth's soil and vegetation cycle, how deep oceans were here, and what kinds of currents they produced. We knew how reflective the planet was across all wavelengths, taking into account Earth's atmosphere and surface combined. This overall "broadband" planetary albedo was a chief input to 1D EBMs. So the fact that we couldn't really simulate a full, dynamic atmosphere interacting with light from different types of stars wasn't that big a deal. But when it came to exoplanets, whose atmospheres and surfaces we knew nothing about, orbiting stars whose light was distributed differently from that of the Sun across all wavelengths, it was a different story. We could do some things with quick 1D models. But to learn more and feel more confident about it, we needed to go to 3D. 3D simulations took a lot longer to run. But they were worth it.

The work was powered by a single question: Could host star spectrum interacting with water ice on a planet's surface and gases in its atmosphere affect the planet's climate and potential to be habitable? In other words, if we took the time to look into what happened as their stars' light hit the air and the ground on these worlds, might there be many, many more potentially habitable planets out there than we think?

In a GCM there are so many knobs to turn, it can be easy to get lost. So we started with the Earth—the one planet we knew to be habitable. And we put the Earth around different types of stars, at the orbital distance required for the planet to receive the same total amount of light from each star. It was *how* that total amount of light was distributed across the electromagnetic spectrum that was different. First, we chose a star that emits a lot more infrared light than the Sun: those small, cool, red M dwarfs. Then we started our simulations. We let the star's light shine down on the planet, some getting absorbed by the atmosphere and surface, and some reflected. We let the simulations run as the planet rotated on its axis and revolved around the star, through seasons and years. Then we checked if the planet was in equilibrium, which meant the average surface temperature didn't change by more than a degree over twenty years of our simulations. And we took a look at the planet's climate. We did this for planets around cooler stars than the Sun, and we did it for planets orbiting hotter stars than the Sun, that emitted more light at visible and UV wavelengths. We compared these planets' climates. This is how we would answer our science question. We turned only the knob of host star spectrum and looked to see if there were any chips to fall.

There were. Turns out the FameLab speech, which was based on work I'd done with less-sophisticated models, was right. Planets orbiting cooler stars were actually warmer than planets receiving the same total amount of light from hotter stars. The type of light a planet received from its star mattered. Carbon dioxide and water vapor in the atmosphere, and water ice on the surface, absorbed the hell out of infrared light, making planets around cooler stars warmer than others. It looked like that knob—host star spectrum—made a difference. Planets around M dwarfs appeared to be harder to freeze into snowball planets than planets around hotter, brighter stars.

This was a result. Indeed, it was the result of the first paper published with my name as the first author. This is a milestone for every graduate student—that first symbol, published in indelible ink, of scientific independence. (Not to leave out the incredible work of a grad student advisor—mine, Vikki, suffered through editing numerous drafts and helped turn a rambling mess into a tight and coherent twenty-five-page journal article. Luckily, I had one of the best.) Before the result, no one knew that host star spectrum interacting with water ice on a planet's surface could have a measurable effect on its climate. It was a significant contribution to the field. It is still my most cited first-author paper.

Later, we found that not only were M-dwarf planets harder to freeze; they were also easier to thaw out of snowball states, if they got themselves into them. Another result. Another paper. Another part of my dissertation completed. I was told that I needed to write three papers to finish my PhD. I was now two-thirds of the way there.

Gravitational Interactions

The final year of my dissertation, one of my professors, Eric Agol, discovered a planet. But it wasn't just any planet. It was one of the smallest potentially habitable worlds we'd found so far, only 40 percent larger than the Earth, or 1.4 times its size. We had learned from statistics that most planets smaller than 1.5 to 1.6 times the size of the Earth had a pretty good chance of being rocky. If they were rocky, then they could have things like oceans and lakes on their surfaces, especially if they were in the habitable zone, which this planet was. It was near the outer edge, so it would need to have a pretty thick atmosphere to stay warm enough for water to stay liquid on the surface. But it was a good prospect for a habitable world.

This planet was named Kepler-62f. It was the outermost planet in a five-planet system found orbiting a K star—a star a little cooler and smaller than the Sun. The fact that this planet had siblings—four at least, and possibly more that hadn't been detected—made this planet particularly special. And complicated. Planets can push and pull on each other, and those interactions can cause their orbits to change. A change in a planet's orbital shape, or the distance from its star, or the tilt of its axis—

GRAVITATIONAL INTERACTIONS

all of these things could impact the planet's ability to host life. I had never looked into the effect on climate and habitability of the kinds of gravitational interactions that could occur among planets. I would need to learn to use different models—models that were designed to study how galaxies formed and merged, and how stars moved within those galaxies. These models were called n-body models, and they'd been used to study the formation and evolutionary history of the solar system. With the help of Rory Barnes, an orbital dynamicist in my department, I learned to use them to study the evolutionary history of another system around a different star. I learned how to figure out where each planet was when one of the other planets was crossing in front of their host star from our viewing angle on the Earth. Then I simulated their orbits for a million years. I ran hundreds of simulations, each time changing the shape of the orbit, and determined how elongated (eccentric) the orbit could get for Kepler-62f without any of its sibling planets getting kicked out of the system, which can happen to planets over time. Then I used that information—the range of eccentricities the planet could have while still maintaining the stability of the entire system—to run climate simulations with my 3D GCM. At the time, no one had done this—used the results from running these dynamical n-body models to inform climate models. We came up with a range of atmospheric compositions, obliquities, and eccentricities under which Kepler-62f could be warm enough for liquid water to flow on its surface. We were able to show that this planet was a pretty good prospect for a habitable world. It was very far away—1,200 light-years, so it was unlikely that we'd be able to follow up on it with the next generation of space telescopes. They wouldn't be sensitive enough to peer into the atmospheres and onto the surfaces of planets that far away yet. But I had established a solid method for vetting potentially habitable planets. If we could do it with this one, then we could use this method to identify habitability requirements for planets that were a lot closer to us, where we had a better chance of checking to see if we were right.

At the same time, during my fifth and final year, I was applying for jobs—postdoctoral fellowships, mainly. I needed to identify faculty at different institutions who were working on what I wanted to do during this next phase of my career, the transient three-year-long phase somewhere in the space between student and faculty. Steven and I weren't thrilled about the possibility of going somewhere temporarily. We felt we were too old to keep moving around. It was time to settle down somewhere, preferably in Southern California near LA, where he could get back to the acting career he'd left behind on behalf of Team Shields. But the reality was that this was part of the deal if I was going to continue as a scientist. I might have been able to skip a postdoc if I'd stayed in theater, but not in science. So I needed to make initial contacts with professors at institutions in cities I was halfway interested in living in for at least a few years. I would need to decide what I wanted to study. What was my big idea for my postdoc? And whose expertise could I leverage to work on it? How was I going to take what I'd already done with the interactions between a star's spectrum and a planet's atmosphere and surface to the next level? And how was I going to get a fellowship at a place in the world where both of us would want to live?

I wanted to understand more about what could influence the climate and habitability of planets orbiting these cool, red M-dwarf stars. I'd met John Johnson—an exoplanet astronomer who built telescopes to find M-dwarf planets—at a conference of the American Astronomical Society, or AAS. He was one of the youngest professors to ever come to Caltech. And he was African American. I'd used my acting marketing skills and sent him a follow-up business card in the mail tucked into a thank-you note. Several months later, he showed up at UW to give a colloquium. Afterward, I went to the regular graduate student dinner with the speaker. During the dinner, John mentioned receiving my card and told everyone at the table that this was what they all should be doing—sending business cards, marketing themselves. I was floored. He later emailed me and told me that he was moving to Harvard with tenure, and that he wanted me to come work with him for my postdoc.

I crafted a proposal for an NSF postdoc fellowship to study M-dwarf planet habitability with John at the Harvard-Smithsonian Center for Astrophysics. I also developed a proposal for a Hubble Fellowship to work with Brad Hansen at UCLA. An orbital dynamicist, Brad would be an excellent advisor to work with on furthering my understanding of the climatic effects of gravitational interactions in M-dwarf planetary systems. And LA was certainly the best place for Steve to be if he wanted to keep pursuing his acting career, which he did.

I didn't get offered the Hubble Fellowship, but I did get the NSF one. A postdoc fellowship. It was every graduate student's dream—a job. But it was not just a job. It was a job at Harvard. Who would turn that down?

I was elated and confused at the same time. Did this mean I was moving to Boston? Would Steven go with me? What about his career? I mentioned my "predicament," if you could call it that (*Oh, I have this job I've been offered at Harvard, woe's me . . .*), in an email to a mentor and brilliant exoplanet astronomer, Phil Muirhead. He put a new idea into my head, as if it had always been there.

"Why don't you ask NSF if you can move the fellowship to UCLA? Tell them you have a two-body problem." (Scientists use this physics metaphor to describe the issue of how to find jobs in the same city for dual-career couples. Yes, we are indeed nerds, in case that was ever in question.) "It's worth asking."

I brought it up with CC, as she'd recently advised some postdocs with unconventional work situations. She said she'd even had a postdoc who spent part of their year in Seattle and the other part of their time in another state. Since most of what we did for our research involved computer modeling, which could be done anywhere with a strong internet connection, this hadn't been a problem for her or for the postdoc. I felt hopeful.

I called the program officer for the NSF postdoc fellowship. I still remember how I felt when she said over the phone, "We want to find the solution that works for your whole life." Even if they never funded me again, I would love the National Science Foundation for the rest of my life.

I knew I wanted the opportunity to work with John at Harvard. But I didn't want to be there the entire year. With CC's help I came up with a few options to present to him, all of which Steven and I had decided in advance we could tolerate. The one John liked best was where I spent the academic year at UCLA and my summers at Harvard. Steven would visit as much as he could during the summer. And during the other nine months of the year, we would live under the same roof. He could continue to pursue his career while holding down a day job, and I could continue to pursue mine. This arrangement might not get me my pick of faculty positions later, if indeed that was the route I wanted to go, which I wasn't altogether convinced of anyway. But it just might help me stay happily married. *Happily* was the operative word. It was about what I could live with, and what I couldn't. I discovered that when I took actions based on my own value system rather than anyone else's, I was a happier person. Being happy was more important to me than being prestigious. I could live with that.

Congratulations, Dr. Shields

My PhD dissertation was titled "The Effect of Star-Planet Interactions on Planetary Climate." The number of planets found orbiting other stars now totaled over 4,300. That's after surveying only the tiniest fraction of our own galaxy. And there are over a hundred billion other galaxies besides our own. Billions of planets orbit distant stars, waiting to be found. About the ones we have discovered, particularly those close in size to the Earth, we know almost nothing. We don't know what gases are in their atmospheres. We don't know if their surfaces are dusty or wet or icy, or a mixture of all three. We don't know what the weather on these worlds is like, or whether life might swim or crawl or walk on their surfaces, or did so long ago but does not now, and why that is.

I'm thinking of that time back when I was a senior at Exeter, when I wrote that meditation called "High Flight" about the journey I took to discovering that I wanted to be an astronaut. When I said out loud toward the end of the piece, "I mean, I'm actually going to do this!" And people cheered, and I was so proud. And then when I went on to MIT, the reality of this field and what it would take to do well set in, and my life took a very different path. Nothing linear about it.

I found my way back and approached my path to the stars from a different angle than I had before. Rather than coming at it directly, like a head-on collision, I sidled up to it again, holding it lighter this time, acknowledging the wide, meandering way that I'd come back, and the possibilities that path had opened up for me. As I stood in front of my department, advisors, husband, mom, brother and sister, best friend and her husband, defending my dissertation in the summer of 2014, I knew that I wanted to continue in astronomy. But in the far corner of my mind I knew that I would never be a typical scientist. I was going to have to do this as the person I had become, not as the person I thought I'd needed to be when I was twelve years old and dreamed up this whole master plan.

Several months before I actually defended my dissertation, I attended my PhD hooding ceremony. Universities can't wait for graduate students to finish writing. They have schedules to maintain. So I showed up dressed in my graduation robes, as is customary. I knew I still had a lot of work to do, and the full celebration would come later. Steven had a 103-degree fever, but he wasn't going to miss it. My father flew in with two video cameras. He wore a dazzling deep-purple suit. It matched the school colors of my robes perfectly. We hadn't coordinated. He is just that fashionable. My advisor was there, and as she placed my PhD hood over my head, chills ran down my spine—all of this no longer the stuff of my wildest dreams. All paths had crossed at this intersection point. And I was awake to it.

Dad also brought something else. A thick amber necklace that his mother—my Nana Doris, a lifelong teacher—had worn. He put it around my neck, on top of my PhD hood. It was the perfect complement. As he snapped a photo, my eyes lit up the frame.

After I defended months later, my mom, brother, and sister stuck around for a few days to visit. I showed them the sights of Seattle, took them to eat dim sum at Din Tai Fung. My brother, the mountain man, convinced me to hike up Little Si, the smaller of two mountains a forty-five-minute drive from downtown Seattle. Little Si is big. In August, during the middle of the single week of the year when Seattle becomes unbearably hot and muggy, and some people actually die because no buildings have air-conditioning, it took several hours to summit its peak. Black flies swarmed around us as we ascended the worn dirt path. When we finally made it to the top, I sat there in a crag between two boulders, peering out across to the Olympic Mountains. I was sweaty and annoyed at my brother for dragging me on this trek. I had been working hard for five years straight. I didn't want to put in any more effort. But deep in my belly there was a warm, satisfied feeling. I had done this. This thing I was sure I would not be able to finish once I started was now in my rearview. After all of that effort, I could now enjoy the view and let myself be cradled in its majesty. Wasn't this how I liked to do things, after all?

When I typed out the last line of my dissertation a few weeks later, and submitted it to UW for permanent archiving, I ran out to the living room where Steven was playing video games. We danced to Pharrell Williams's "Happy." Steven isn't a natural dancer, but he stayed with me, shaking out every mile, lap, and tear of the last five years, for the length of the entire song. Two days later we were in Victoria, British Columbia. I hadn't been there since I was nine years old. We stayed at a luxury resort in Oak Bay and lounged by the pool overlooking the ocean. We watched Matthew McConaughey in a private hotel screening of *Mud* with individual popcorn and candy servings. We were the only couple there under fifty. We ventured off the resort grounds for three occasions: tea at the Empress Hotel; a visit to Butchart Gardens; and to find cheese scones, which I'd remembered falling in love with as a child. We found them, and as I sank my teeth into one, warm and filled with delicate bits of yellow cheddar, the butter I'd smeared into the dough soaking into the nooks of air in the scone, I was happy. After our long marathon, this was just my speed.

Five years earlier, on the first day of the UW orientation for women who were reentering school after a significant time away, the invited speaker, herself an older returning student, told us to go buy a graduation card. She said to fill it out "Congratulations, Dr. _____," and keep it with us. Anytime school got overwhelming, we could pull out that card and look at it and know that if we kept at it, one day that card would be true—we would be Dr. _____.

At the urging of Nancy Finelli, the UW reentry program director, who became one of my mentors, I did this. I went to the campus bookstore and stood in front of the rows of graduation cards, flipping open card after card until I found one with a quote from Confucius: "The will to win, the desire to succeed, the urge to reach your full potential—these are the keys that will unlock the door to personal excellence."

I paid for the card, brought it home, and, sitting at my desk with a pen hovering over the inside cover, I placed the ballpoint onto the cardstock and wrote as if I were signing a card to a friend: "Congratulations, Dr. Shields. PhD in Astronomy and Astrobiology. You did it!"

I carried around this card in a folder in my schoolbag, along with a vision collage of the kind of life I wanted to have as a scientist. I didn't have a detailed explanation for why I'd chosen the images that were there. For once, I hadn't thought. I'd just torn pictures out of magazines, cut, glued, and pasted. There were pictures of Jupiter's moon Europa. There was a picture of a space shuttle. There were blown-up images of microbes. There was my own handwriting, mentioning being a generous scientist and a powerful communicator and a fantastic wife.

Over those five years of graduate school I filled this folder with other things: a copy of the email I received informing me that my first "first-author" paper had been accepted for publication in a peer-reviewed journal;

copies of the panel reviews of the successful grant proposal I submitted for a Graduate Research Fellowship from the National Science Foundation; informational pages about stress relief in academia. This glossy purple folder, with the University of Washington seal on the front, frayed corners, and bulging envelopes, soothed me over and over again. As a postdoc, I put the graduation card next to my computer in my office at UCLA. As I typed away on new papers below my framed PhD diploma on the wall, there it sat, reminding me that hard work and perseverance through the flames of fear, doubt, and confusion pay off.

MERGING

Different Stars

Starting my postdoctoral appointment was like beginning to walk slowly after a 10,000-meter track race at full speed. I had been pushing so hard for so long, and I hadn't yet come down from it all. The feeling of frenetic motion persisted. I was still living the life of a graduate student. I worked six days a week. When I was with my husband, I was thinking about work and all that was still undone. A third of the way into my first year, I had gotten sick three times in as many months. It was unsustainable.

I met a dean at UCLA where I was working, and she told me that the only thing I had to do was get enough sleep. In the space between her words, I saw rays of light shooting out through clouds. I remembered the words of another mentor, astronomer Sarah Ballard, telling me that this year was about recovering from the dissertation year. Really, she told me, the most I should expect of myself was finding out how to get a parking pass on campus.

After five years we were back in LA. Everything felt different. Instead of living in a cramped one-bedroom apartment off Sunset Boulevard in the middle of everything, we were now renting a large, two-bedroom multilevel condo in the Valley from one of our former acting instructors. Our two cats, Mama and Mimi, the mother and daughter duo we had adopted a few years before moving to Seattle, had made it back to LA with us, just barely, nearly overheating in our air-conditioning–less car on the trip. They had taken to the extra space in our home.

I was working in the physics and astronomy department at UCLA, in the central part of campus. This seemed like a foreign country compared to north campus, where the school of theater, film, and television was, adjacent to the sculpture garden, where we'd sit outside and eat lunch and sometimes have movement class or rehearsals. The memory of the acting life at UCLA seemed like a faded photograph of someone else.

My parking pass in hand, I began to work slowly on finishing up the Kepler-62f paper for submission, while arranging the first of my creative arts–based astronomy workshops for middle school girls in LA. NSF also cared about the broader impacts of a scientist's research, so I'd needed to come up with an educational outreach plan to go along with my research plan for the next three years.

Where are the other faces like mine? I wondered. It wasn't that women of color just weren't as interested in astronomy as white women or men, which is the story that people who don't want to ask this question tell themselves and everyone else who cares about diversity and inclusion. It was something else.

I had occasionally done outreach programs with girls of color while I was in Seattle. I'd invite a group of middle or high school girls from a neighboring school to the planetarium on campus, give them a star show, and then play theater games. They seemed to enjoy engaging in something interactive. They weren't worried about getting something wrong. It wasn't a test. They relaxed in their seats as I showed them how to find the Big

Dipper and the North Star and the Little Dipper. I sent them home with a star chart to use when they looked up at the sky on a clear night, and they seemed happy.

I went back to my own story. I thought I had to become something that I wasn't in order to be a scientist. I loved magazines and makeup, fashion, and being a girl. I had parents who were musicians, a grandmother who loved math but had been ignored in college, never called on when she raised her hand in class. This history was part of me. How could I not bring it to the study of what I loved most in the world?

I wanted to find a way to encourage young girls to bring their whole selves to what they learned. The creative arts—theater, writing, and visual art—are by their very natures subjective. Everyone views art differently, bringing their personal likes, dislikes, and biases to bear on their opinions about a piece of art. How someone acts, writes, draws—these are all informed by the unique individual, who that person is and where that person came from. The legacy of that person's development and evolution on this planet is like a fingerprint on a piece of art in the world. I wanted these girls to put their personal fingerprints on what they learned about the universe. My gateway to getting them to do that was to come at science through the inherently personal, subjective world of the arts.

I decided to propose creating a four- to six-week workshop called "Universe: More Than Meets the Eye" for girls from groups traditionally underrepresented in the sciences. In the United States that meant girls who identified as Indigenous American or Alaska Native, African American, Hispanic or Latina, or Native Hawaiian. I would focus on middle-school girls, ages ten to fourteen. The education literature said that this was the age when girls started to get quiet, to focus more on their physical appearance and less on how they think and feel about the world. They stop raising their hands in class because they are afraid of sounding stupid. I knew this feeling well. I wanted these girls to see that, just as there is so much information we would miss out on if we only looked at the disk of our Milky Way galaxy in visible light, rather than also looking at the infrared and X-ray light it emits, so too was there so much more to these girls than what was on the outside.

I would teach these girls astronomy and astrobiology, but I would do it using theater, writing, and visual art. They wouldn't just learn what a star or a planet was. They would write a poem about that star or planet. They would paint a picture, calculate the distance to Saturn in units of themselves, rather than kilometers or miles. They would learn that constellations are merely patterns in the sky, and they could create their own patterns and constellations and name them, and write myths for them just like the Greeks or African tribes or Native American nations had done. We would do it this way so that I wasn't just filling their heads with information that could pour right out with no personal connection to it. Instead, when they wrote or drew or danced or sang something related to what they were learning about the universe, they would, I hoped, feel connected to that astronomical phenomenon. They could look up into the sky at a star and say, "That's mine. I wrote a poem about it." And maybe when all the heavy math came in as they continued to study the night sky, they wouldn't run away or think that they didn't belong. Maybe encouraging their personal ownership of the universe, of which they were an integral part, would help them stay.

I walk into a classroom of Brown faces in a middle school in Eagle Rock. All of the girls are Hispanic/Latina. They are busy. They don't want to be here. They'd rather be working on homework. I've hijacked their after-school study hall twice a week, while they wait for their parents to get off from work and come pick them up.

I start to talk. I tell them about myself. I show them pictures of my mom and dad on my laptop. I show them a picture of me when I was nine years old and watch their eyes get big. They seem surprised to see that I was ever little. To them, a teacher is born fully grown out of the womb. Now they believe something else.

I pass around composition books that I call "playbooks" instead of workbooks, hoping that will help. I spread out magazines—*Astronomy, Sky & Telescope*—full of glossy pictures of galaxies and star clusters and nebulae. I tell them to start ripping. They are excited by this. I tell them to choose pictures they like and glue them to the front and back covers of their books. They create books covered in rockets and the northern lights and the moons of Jupiter and the blues of the ice giants—Neptune, Uranus. Some of them ask me more about what is in the pictures they've chosen as they're gluing them onto their books. I tell them a few things and say we'll get to it all later. For now, I say, make these books your own. I tell them I'll collect them at the end of each meeting so they won't have to worry about bringing (or forgetting) them next time, but at the end of the workshop they'll get to keep them.

I ask them to write for five minutes in their playbooks: "When I think of the word *astronomy*, I think of . . ." and fill in the blank. Later, when I read through their writing, I see words like *stars, planets, the universe*. I ask them to do the same thing for the word *astrobiology*. A few write "life." Most write that they don't know what the word means. It makes sense. The word is only a few decades old, the field less than fifty, compared with astronomy, which has been around for thousands of years. By the end of the workshop they will know what both words mean. After the calculation Aitana does of the distance from the Earth to Saturn—

one trillion Aitanas from head to foot; after Kira tells me emphatically, almost angrily, that she doesn't want to calculate her distance to Pluto because "it isn't even a planet!"; after I watch them all draw their imagined landscapes of an extrasolar planet and see in some pictures three stars suspended in the sky—a triple star system; after I read in their playbooks weeks later, "When I think of astrobiology, I think of life on other planets, aliens, ET" . . . I smile.

Harvard

After spending my first academic year as a postdoc in LA with Steven, it was time for me to hold up my end of the bargain and go to Harvard for the summer. Steve and I planned out visits—three of them over the course of ten weeks. Having them on the books made this thing—the reality of being apart—less scary. It could be like the early days of dating if we let it. A long-distance romance. A distance-making-the-heart-grow-fonder kind of thing. We hoped for the best, and I got on a plane.

I'd lined up a sublet for a studio apartment of a Harvard Law student. It was tiny but clean and well furnished, with small appliances and a small couch and a small bed, everything fitting in its place. And it was a short walk from Harvard Square.

It was intimidating walking into the Harvard-Smithsonian Center for Astrophysics for the first time, the "CfA" as it's called. Its older red-brick building with burnished metal telescope domes stood out against its more modern extensions of glass. This was the place where Annie Jump Cannon classified stars in the late nineteenth to early twentieth century. We still use her system today. The Harvard women "computers" were a legion of white women who measured and cataloged astronomical phenomena—stars, galaxies, nebulae—yet were given limited opportunities for recognition of their efforts. Henrietta Swan Leavitt was hired to work on variable stars but was not allowed to earn a degree from Harvard. Her degree was issued from Radcliffe College, Harvard's female affiliate. Much later, Cecilia Payne-Gaposchkin would earn the first doctorate in astronomy from Radcliffe, in 1925. The two colleges wouldn't merge admissions until 1975; "sex-blind" admissions commenced two years after that. And Black women? Alberta Scott graduated from Radcliffe in 1898. Ayana Holloway Arce received her PhD in experimental high energy physics from Harvard in 2006. Candice Etson received hers in biophysics in 2010. Barbara Williams was the first African American woman to receive a PhD in astronomy ever, from the University of Maryland at College Park in 1981. Jamila Pegues became the first Black woman to receive a PhD in astronomy from Harvard in 2021. At the time, according to *Nature Astronomy*, she was the twenty-first Black woman to ever earn a doctorate in astronomy/astrophysics in the United States. When I received my PhD from the University of Washington in 2014, according to the list compiled by the organization African American Women in Physics, I was the fifteenth. If we include African American women who have earned doctorates in other physics-related disciplines, the list is a little longer, currently about 160 to date. This figure is about the same as the number of students (all races) in the United States who earn doctorates in astronomy each year.

I found the meeting room of the summer cohort of Banneker Institute students. This was John Johnson's program. It was something he'd envisioned long ago. Harvard had brought him here, with tenure, from Caltech, with an agreement to provide funds to create a program to prepare Black and Brown undergraduates for graduate school in astronomy and astrophysics. They took courses over the summer. They would do research, but what set this program apart from the rest was that they also would be prepared mentally and emotionally to undertake one of the greatest challenges of their lives—thriving in departments set up to exclude them. They would be trained in the history of systemic racism, and in social justice and activism, so they would be able to navigate these white environments. They would be strong, they would be full, and they would be free to be themselves. They would be ready. The program was named after eighteenth-century African American astronomer and mathematician Benjamin Banneker. I was here to do research on my own, and to teach the Banneker students science communication.

I used theater games and improv exercises. I thought of activities that combined both science and theater and felt organic to me. One of them I called "Scicomm Ping-Pong." I had students begin by talking about a positive memory they had of a time spent with family or friends. They would start to tell the class the story of this time. Then I rang a bell, and they had to switch to talking about some aspect of their science. I rang the bell again, and they switched back to telling the personal story. Then they switched again. Then again.

The students responded to the exercise. When I asked them why they thought I'd had them do it, a twenty-one-year-old student from Ecuador said, "I noticed that I talk a certain way when I talk about science. And I am much more relaxed when I talk about my friends and family." I wanted them to one day be able to talk about science the way they talked about their families—with warmth, relaxation, and a big world of energy filling their bodies. This was one way to begin that process.

I floated around in this environment. John, a short, stocky powerhouse and natural leader, was a god to these students. An astronomer. A professor. Tenured. And Black. At Harvard. And yet, so down-to-earth. He walked around in short-sleeved alligator shirts and jeans. He had achieved the highest status someone in our field could achieve. And he had done it while being Black. He had truly arrived. He could relax.

And yet, years later he would share with me how he'd kept going and going, like the Energizer Bunny, as if he was still on the tenure track, as if he hadn't attained that revered status of permanent job security and eternal prestige. No one told him he didn't have to work that hard anymore. No one would ever tell him that. It was he, finally, who had to be the one to tell that to himself, and act accordingly.

I'd spent the better part of the first year at UCLA finishing up the paper on Kepler-62f. I had been trying to get a new paper off the ground with Brad Hansen, my UCLA advisor. I spent that first summer in fits and starts with it, sharing small steps forward at John's group meeting and developing the seeds of a course to teach scientists how to talk about their science as if it were the great love of their life and the most exciting adventure story, enough to make Agatha Christie and Indiana Jones salivate.

Boston in the summer is magical. The Charles River glistens with light, and the oars of sculling boats splash and glide through it smoothly and gently, like skating on ice without sound. I walked back and forth to work, through the bustle of Harvard Square in Cambridge, footsteps from MIT, where I had been a lifetime ago. I always thought Harvard probably would have been a more logical place for me to have gone to undergrad, but I never even applied. I had stars in my eyes for MIT alone. Now that I was there at last, I felt ancient and new at the same time.

When Steven visited, my time there came alive. I could show him everything I had discovered while we were apart. We devoured cannoli at Mike's Pastry, ate pizza and pasta in the Italian North End, and walked the Freedom Trail, getting caught in the rain. He folded himself into my tiny apartment, burying deep under the covers with me in the small bed, and went to barbecues at John's house with the Banneker students. Then, like magic, he was gone, and I walked the cobblestone streets of Cambridge alone, trying to be brave and independent.

The Banneker students were a mirror for me. There I was, this example of where you could go as a Black woman—a Black *person*—in astronomy. There were several of us postdocs there to be a part of what John had created, and he made sure we got the respect he felt we deserved. I was to be called "Dr. Shields." Until students themselves had obtained a PhD, they called those who had one "doctor." After that, a first-name basis was acceptable. I made a note to continue that tradition when I got my own research group together.

But I felt twinges of the feelings I'd had at Exeter looking across the dining room at the Afro-Latino Exonian Society table. I wasn't used to being around large groups of Black students in astronomy environments. Why did I feel less comfortable in the beginning than when I was alone in a sea of white astronomy students and faculty? This went back to my childhood, living in graduate student housing in La Jolla, in Canadian and northeastern U.S. suburbs—everywhere where everyone was mostly white. I felt safer among what I knew and saw every day than around people who had skin the same color as mine. I held my breath when a Black man walked by me on the street. As I grew older, I forced myself not to grab my purse a little tighter, not to pick up my pace. I made eye contact and smiled at him through my fear. I reminded myself that serial killers were predominantly white and male, not Black. I felt twisted and turned around, and while I knew that my kindred spirits, the people who "got" me, crossed color lines—from Jonna and Shelly to Yvonne, Steven, and my white girlfriends Ashley, Denise, and Kara—I wanted to feel at home with Black people everywhere. I wanted to feel that they were "my people" too, just because they were Black, like me, with ancestors who were enslaved, like my great-great-great-grandma Cole. That alone was enough to bond us all together. I knew that I should carry the freedom to dislike a Black person for being an asshole, just like a white person has the freedom to love and hate other people who are white and jackasses. But first, I wanted to claim all Black people as my kindred. I wanted them to start out ahead of the pack in my heart, rather than behind. I felt this was only right.

Maybe I was afraid of once again being judged, not fitting in, not being seen as "Black enough." Even with my senior status among these budding young Black and Brown undergraduate juniors, I wanted to be accepted as someone they looked up to, not only as an astronomer who could do good science and publish papers. I also wanted them to look up to me as someone who had done it while holding on to my heritage and to who I was. I knew that those two things weren't necessarily one and the same. So here I was again, wanting to be validated.

But I was their instructor, so those feelings got pushed to the side. I let them come out and play on Sundays when John had the barbecues at his house. Sometimes the feelings traveled the blocks with me there. Sometimes they kept me home.

John and I needed to figure out a paper to work on together. We walked to his favorite Chinese restaurant for lunch, and over hot and sour soup, he brought up an idea. No one had written a recent review paper about the prospects for habitability on planets orbiting M-dwarf stars.

These small red stars were the most numerous stars in the galaxy, making up 70 percent of all stars. Life is likelier to exist on a planet orbiting this kind of star due to the sheer number of them alone. Red or M-dwarf stars are also easier targets for the current ways we look for new planets, by watching for telltale signs that something is orbiting and maybe even crossing in front of a star from our viewpoint.

These stars are extremely long-lived. They are the long-distance runners, conserving their energy so they can last the whole race—the tortoises of the stellar family. Remember when I told you that mass means everything when it comes to the future of a star? Those high-mass white and blue stars I talked about earlier run through their fuel in a flash. But these red, low-mass stars burn through their fuel very slowly, living for billions, and in some cases trillions, of years. We know how long they'll live because of their mass. Mass tells us how much fuel these stars have in them. No red stars have yet died in the entire existence of the universe, which is about 13.8 billion years old. By comparison, life took around half a billion to a billion years to get a foothold on Earth. The exceptionally long life spans of M-dwarf stars may allow more time for life on orbiting planets to develop.

Or maybe not. These stars are tricky, and other considerations could complicate things.

Planets in the habitable zones of M-dwarf stars are much closer to their stars than Earth is to the Sun. When planets get close to their stars, the stars can slow down their rotation rates. We call this "tidal locking." The most extreme case of this is synchronous rotation, where the planet takes as long to turn once on its axis as it does to orbit once around its star. A synchronously rotating planet would always show the same side to the Sun—the day side—while the other side would always face away, and it would always be night. I mentioned this type of planet at the very beginning of the story. Remember?

Habitable-zone M-dwarf planets could be synchronously rotating. But could they support life if one side roasted while on the other side it got so cold that the atmosphere froze out onto the surface? Recent work theorizes that with enough greenhouse gases in its atmosphere, a synchronously rotating planet could be capable of pulling enough heat around to the night side to keep this from happening and to even out the temperature contrasts between the two sides. And, of course, there was the new scenario my team had proposed and proven possible for this kind of planet, where life—if present on such a world—could find a haven along the terminator dividing the two insanely unappealing halves. This was the power of theoretical modeling. We could show that something we couldn't yet measure or even detect on a planet could make a big difference when it came to the potential of that planet to harbor conditions amenable enough for life.

There were other issues. Because M-dwarf stars are so long-lived, they behave like toddlers in their terrible-twos phase, wildly spewing out excessive amounts of high-energy light toward their planets. X-ray and UV radiation is bad for biology. On the Earth we wear sunblock to protect us from UV rays. All that XUV light pelting the planet's surface could make it difficult, if not downright impossible, for life to survive anywhere else but in the deep ocean—if oceans exist on these worlds.

Life is life, so who cares where it resides, right? Actually, *I* do. If that life doesn't live on the planet's surface, then from our standpoint, it might as well not even be there at all. As one of my professors used to say, the ability of life to exist on a planet's surface gives it the best possible chance of making its presence known in ways that we might be able to observe remotely, from space. It could excrete something—a gas, for example, like the carbon dioxide that we exhale. And with a sensitive enough telescope, we might be able to measure that, and it could tell us that life is there. But that would be tricky too. A lot of things can excrete carbon dioxide. Volcanoes do it too. So figuring out the telltale recipe of what to look for, that only life can produce, is another task that some scientist somewhere is working on right now.

The last review papers on this subject had been written in 2007, nearly a decade earlier. In that time, so much had changed. We had the Kepler and K2 missions, which had found thousands of planets, many orbiting these cool, red stars. New ground-based telescopes had been built by John and others, and these platforms had found more planets around these stars. We had done so much theoretical work to understand how the unique stellar environments of these stars could impact planetary habitability. There were literally hundreds of original research papers that had been published on these topics over the previous eight years. The major contributions, and how they directed our understanding of these stars as hosts for habitable planets, needed to be summarized in one place. John wanted to know if I was interested in taking on that task. It was strategic. I would learn a lot, to be sure. I would write a review paper that was likely to be highly cited. And I would establish myself as the go-to person when it came to the habitability of M-dwarf planets. In short, it would get me a faculty job.

TED

Sometimes galaxies collide. Black holes merge to form new, misshapen phenomena never anticipated or dreamed of before.

I had no role model who had done what I was trying to do. I had devoted five years to astronomy alone, letting go of the creative arts side of me, apart from the odd art class or poem written and shoved hastily in an envelope to a journal somewhere in the world. But now I wanted both. I wanted to use my acting background to become a better scientist. I had terminal degrees in both acting and astronomy. This could work, but I would have to create the role I wanted from scratch.

I didn't see it coming. I was good at conforming, trying to fit in, making it work. It saw me coming and swooped in.

I got an email from Lucianne Walkowicz, a stellar astrophysicist and artist with jet-black hair and blue-green eyes. Lucianne wore makeup, like me, and used stellar pulsations to create the music of the universe. What I didn't know, what this email was about, was that there was a program that helped people like Lucianne and other game changers and global innovators find their people. Applications to be new TED Fellows were due the next day, and Lucianne hoped that I would apply.

I dashed off the application and clicked SUBMIT. Then months went by, and I forgot about the whole thing.

I remember the phone interview with the program's director. I explained something about planets to him. Then I heard myself say, "Did I lose you?" and I tried again. I think maybe that's what got him. Not all scientists care if they lose their audience.

As a TED Fellow, I got to give a five-minute TED Talk on the TED stage, in Vancouver, British Columbia, where the TED conference is held every year. Five minutes to share your "big idea," to show people who you are and explain why they shouldn't forget.

I was defiant during the first video call with the TED Fellows team where I shared a draft of my talk for the first time. I didn't want this to just be about science. That's not who I was. I wanted the world to know that I was a scientist, yes, doing amazing science, yes. But I was so much more. I was not your typical scientist. I knew the sciences and the arts could fit together, and I wanted everyone to know about it. TED had encouraged me to launch my big idea in my talk. I had finally given it a name.

Stars shine in many colors. There are yellow stars, like our Sun. There are orange stars and white stars. And there are blue stars and red stars. The most massive stars are the bluish stars. They're the opposite of the small, red stars. Big, blue stars are the sprinters, giving everything they've got for a short amount of time. They don't live for very long, only a few million years. That's not a long time for stars, and not long enough to give life time to develop on an orbiting planet.

My mom called me her "star girl" when I was little. I would call the organization that had grown out of the work from my postdoc—introducing middle school girls of color to their universe using the creative arts—Rising Stargirls. These girls, like the stars they would learn about and come to know, shine in many colors. We would work to help them feel how much they were, truly, "star girls," whether they chose to stay in astronomy or not.

The Fellows team wanted me to lead with the science, to establish my street cred first. I wanted to lead with the personal: how I was different, how I didn't fit in. It was the story of my life. Always wanting to do it differently, always running away from the science, which somehow still felt foreign, comprising a tiny percentage of my cellular makeup. I pouted for a week, and then decided to try it their way.

They were right. I didn't have to beat people over the head with my uniqueness. Once I got their interest and respect with the science, I could drop it in—how this science they'd just learned came from someone with a surprising background. It was a trick that I played on them. *Guess what? I'm an actor too. Bet you didn't know that . . .*

I hadn't planned on getting teary at the end when I talked about the girls I hoped to serve through Rising Stargirls. It just happened. It was the "moment before," like Salome always said. Every second of my life had led up to this moment, and I remembered being that little girl with a big dream in the sky who wanted to be a part of it all. I was glad I had all the voice and speech work underneath me, holding me up. It helped me breathe through the tears so the audience could still hear me. Training is good.

Kids

For seventeen years I had taken a small blue pill every night before bed. I was sure. This was the way life was, and it was fine. Great, actually. Before that, Steven and I had almost broken up less than a year into dating. I was certain that I wanted to have children eventually, and he was certain that he didn't . . . ever. For two hours we argued and cried in each other's arms, sprawled on the bed in his small studio apartment on a side street deep in Hollywood next to what would become the Hollywood & Highland Center, and then later, Ovation Hollywood, home of the Dolby Theatre, where the Oscars are held each year. Then, on our way to rehearsal for the play *Unidentified Human Remains and the True Nature of Love*, we agreed that maybe we didn't need to decide whether we stayed together that day. Maybe we didn't need to have this heavy, serious conversation less than a year after we started dating. Maybe we could just chill the fuck out and give it a rest. So we did. Life went on. We continued to learn how to love each other through talking about and doing the little things, without worrying so much about the big ones.

When I was in grad school for astronomy, up in Seattle, the subject came up again. We'd been married for five years at that point. Did we want to have kids? I certainly didn't want to have them then, while I was too busy as it was, trying to keep my head above water with courses, papers, and the full-time job of trying to prove myself, plus be a good wife. And daughter from afar. And sister. Plus, we were balancing on a grad student salary, and Steve's unemployment from California, followed by relatively meager wages as a ticket seller in the nonunion environment of the Seattle theater scene. We weren't financially stable enough to add a dependent to our household. We could barely afford our cats. So we tabled it until a time when things would seem more stable and permanent.

Then the University of California, Irvine, came knocking, and before I knew it, a job talk had become a job offer. Negotiations led to a mutual agreement, and the potential for fairly stable income for the duration of my working years, if I wanted it, was a reality. And just like that, the financial stability argument was no longer something that held any water.

I was faced with making an actual decision about whether I wanted to have children. I couldn't put it off onto anyone else or any outside circumstances any longer. I needed to make a choice.

Steven and I sat in a Tender Greens booth on the corner of Hollywood and Vine, a block away from our favorite movie theater, the ArcLight, where we'd gone throughout our dating years and into marriage. It was one of the few places, along with our beloved Ethiopian restaurant Meals by Genet, that we missed dearly while in Seattle for those five years of grad school. We sat in that booth, waiting to walk over to see a movie, and we talked about children. The subject had come up now and again since that argument long ago, and way too early, that almost broke us up. Mostly we'd whisper to each other at holiday parties full of friends with kids bouncing off the walls about how glad we were that we weren't our friends. We'd be on an airplane five rows ahead of some wailing toddler and thank our lucky stars that we weren't that toddler's parents, or the unfortunate passenger sitting next to that family. Now we were out to dinner and a movie, a fancy-free childless couple—another opportunity to pat ourselves on the backs for our intelligence and consistent use of birth control.

Steven looked over at a neighboring table, where two parents nearing middle age sat with a toddler in a high chair.

"Look at those two. We'll be their age or older by the time our kid is that age. I'll be close to seventy by the time they're in college."

I noticed how I felt when I heard him say that. Something was different. The feeling was distinct, and unmistakable. It was sadness. Okay, so when I heard him mention the drawbacks of having children in our forties and fifties, it made me sad. What did that mean? Had I come full circle? Or something in between? From an adamant *Yes, I want kids!* to a *Thank God I don't have kids!* to a *Well, maybe I might want kids after all . . . ?*

"I don't know one way or the other whether I want to have children myself," I said to him. "I can think of tons of cons. Not being able to do whatever I want to do, when I want to do it. Not being able to travel frequently. But what if there is a person who is supposed to be here, who we're keeping from existing?"

That night I was able to put some sequence of words together that approximated a position, or at least a dim inclination on the matter. I decided that I felt pulled to stop actively preventing a child from becoming. I wanted to relinquish my turn at the wheel and put a power greater than myself in the driver's seat. Of course, this power had always been in charge to begin with. I was simply finally willing to stop pushing and shoving for control.

☽

In the days and weeks that followed there was more discussion about the cons of having children. No sleep. Having to deal with taking care of another human being for the next eighteen years, minimum. Play dates. Sports lessons, dance lessons, music lessons, carpools. Parent-teacher conferences. Clothes shopping. School shopping. Never being able to get on an airplane, because, Steve told me, we would drive everywhere or not go at all. We talked about the people we had no desire to be. We would not become one of those miserable sets of parents with a screaming child in a floating tin can for five hours on some cross-country trip to see family at the holidays. No way. Nor would we become that couple that brought their kid to a five-star restaurant only for them to scream and wail and throw food and attempt to catapult themselves from their high chair, forcing us to exit the restaurant early to get this monster child out into the air and away from the disapproving, annoyed glares of normal onlookers who had the intelligence to either hire babysitters or avoid the entire enterprise altogether. That would not be us.

Finally, when all the drawbacks of the now seemingly ludicrous idea of even considering having kids had been laid out, in full relief, polished and shiny in neon varnish, I sighed. My sigh was long and thoughtful. Again, the scientist made an appearance in this conversation and observed. I felt—sad again. Depressed, disappointed. I mean, all of these cons, real and legitimate concerns, would be the perfect opportunity for me to have had some kind of lightbulb moment where I went, "Oh my God! It would be a complete drag to have a kid! There would be so much fucking stuff to do! My life is busy enough as it is, what the hell was I thinking? Case closed." Steve and I could finish dinner, pay the check, and continue to the movie portion of our date night—something else we would have a lot more trouble maintaining once we had kids.

Instead, there I was feeling sad about all of the sucky aspects of being a parent. Little, if any, sex. Having to put another human first in all things. Not once had we discussed any advantages—any pros—to having kids.

So I told Steven, "Your job is to find one positive thing about us having children." The drawbacks were so obvious. And my career ambitions and PhD work overwhelm had made it easy to remain tone deaf to any *I must have children* biological clock that might have been going off—the one everyone talks about. I'd wondered what was wrong with me that nearly everyone I knew from high school and college was well into their second kid or third kid or more, and somehow I hadn't gotten the memo that it was time to have kids. But perhaps I had chosen to ignore it. Something had changed between that fight with Steve all those years ago and now. That something was that I had learned about—dare I say, discovered—myself, and what was important to me, in each moment, each year. And I'd gone after that with all my heart, the way many might have gone after producing a child. The thing was, now I was finally taking the time to ask myself if a child was something that I wanted to go after too. And though I couldn't answer that definitively yet, I had become convinced of something: If a child of Steve's and mine was meant to be on this Earth, I wanted to let that happen. I no longer wanted to stand in the way. It wasn't that I felt we needed to have a child to make us a family. Steve, me, the cats—we were already a family. But if it was God's will for our family to expand—well, I wanted to show up for that and leave a place at the table. (And we'll get back to the God thing later, so don't freak out about it now.)

A few days later we were in another restaurant. Between the bruschetta appetizer and my butternut squash ravioli, Steve blurted out, "I've thought of a positive thing about having a child."

I looked up from a bowl of browned butter. "Uh-huh?" I waited.

"It would be us. No matter what else happens, this child would come from us. You and me."

I smiled.

Exploding Stars

There's a difference in the light at the height of summer. The Sun is higher in the sky, brighter, brasher, and that light fills the big blue sky and touches everything. It's as if the world has opened up like a clamshell, and the air is full of sunlight. The bold light blinds you, and all you can do is lie down on the beach or the concrete next to the pool and surrender and let it bake you until you are done.

The second summer at Harvard I lived in Boston proper, right down the street from Fenway Park. It was exciting. Hordes of devoted Red Sox fans flooded the streets whenever the team was at home. Steve and I went to games when he visited and walked across the street to see the sequel to *Independence Day*, about bad aliens returning to Earth to wreak havoc on the human race. There was an art supply store next door, the Museum of Fine Arts a short walk away, and restaurants full of Greek pitas and Southern barbecue with thick, fluffy biscuits around the corner. I lived above an ice-cream parlor. Once again, I had chosen my sublet well.

This was our first summer apart since we'd decided to stop actively preventing pregnancy, a decision that had quickly turned into actively trying to get pregnant. But this was going to be a difficult thing to do while I was living 3,000 miles away. Steve's visits weren't always timed with my ovulation cycle. So we accepted that the summer was likely going to constitute a hiatus from our quest. When Steven was here to devote some time to our endeavor, he would stare right at my lower abdomen after we made love and give it the firm command to "make a baby!" It always made me laugh. While he was in LA, I worked on the review paper, used my newcomer unlimited yoga pass at a studio within walking distance, and ate lots of pita wraps. I taught science communication to another cohort of Banneker students and sat in on their social justice education. Words like *redlining* and *white supremacy* filled my ears for the first time. I worried less about the fact that I hadn't known what those words truly meant until now, or whether the students thought I was the greatest example of Black, female, astronomer, and "woke." I couldn't know everything, and I definitely couldn't be everything. I was learning now, and I was grateful, though it made me sad. But I also felt relief. No wonder it had seemed so hard the first time around in astronomy. It *was* hard. And for the first time, I allowed the possibility that it hadn't been entirely my fault.

In the fall of 2015 the field of astronomy exploded. Not because of more planets being found, or the discovery of the gravitational waves Einstein predicted to exist five decades ago. It exploded—or rather imploded—due to the revelation that one of the most influential and renowned scientists in the field had been sexually harassing female students and postdoctoral researchers for over a decade. He mentored hundreds of students, had tenure, and was largely untouchable. As word continued to spread of the allegations, the female survivors of this abuse continued to pour out of the woodwork, at different institutions, from different decades. I knew several of them. It was a legacy of tragedy.

The tragedy, in my mind, went far beyond the individual experiences of these women, which were horrific in and of themselves. The ultimate tragedy lay in the collective multitude of women who had left the field because of what had happened to them. These young scientists had placed their trust in their advisor to shape their fortunes in this field, counsel them, and support their emergence as exoplanet researchers. Instead, they had been made to feel as if they had nothing to contribute to the field, or that any scientific contributions they would make and any scholarly achievements were not worth the pain and torture they would have to endure to stay in the field. I thought of all the brilliant minds, out there somewhere, cut down, betrayed, stunted, alone, and not putting their brilliance to work advancing the field the way they could have—what their path would have been, had it not intersected that of someone else. It made me sick.

I have yet to be subjected to sexual harassment. I hope I never am, and I am coming to understand now how truly atypical it is not to be, compared to most women in the professional world, unfortunately. And yet I too have felt alone, not belonging to the field of astronomy, and not included. These feelings strangled me early on, when I looked to those in positions of authority to determine the course of my career and my life with their approval or rejection and criticism. When I finally came back to this field many years later, it was not because someone told me I should become an astronomer. No one ever told me that I had a "gift for science." I came back to astronomy because I missed it, and I loved it, and in the interval when I'd gone away, studied acting, and become an actor, the field of astronomy had massively expanded; this new field of exoplanets had emerged, and I wanted to be a part of it. When I surrounded myself with people who supported me, whom I could be vulnerable with and who would not use my vulnerability against me, I was able to have the space to listen to myself, rather than someone else and their opinion about what I should do with my life. When I returned to astronomy, in spite of my fears, I became a stronger person.

And yet, despite my love for this work, the desire to remain in the field, like the Moon, has waxed and waned on a seasonal, if not monthly, basis. If I'd been sexually harassed on top of dealing with my already howling impostor thoughts and all-around fatigue over having to continually be productive and accomplished to stay viable and competitive in this field, I would probably be long gone.

In my mind, I speak directly to the women who have experienced all of it, all of the suffering, across space and time. What I say is half prayer, half plea.

I beg you not to look to others for the answer about whether you belong in this field or that field. The only answer to that question lies within you. It has always been there. People will say anything and do all manner of things, and some are incredibly, irrevocably harmful. With all your might, do not let those things take you away from what you want in this life. It is the only one you get. Live it. Don't let someone else live it for you. Get the help you need, cry, scream, yell, and enlist your armies. Advance on the gates that seek to keep you out. Lean against them with your full weight. Call out for reinforcements. Do not be moved. Do not bend, when it comes to your future. Own it. Stand firm in it. Let absolutely nothing hold you back.

A Bird's Wing in Taos

There is a reason we use the analogy about dry grains of sand when talking about the vast scale of the observable universe. Few people have had the opportunity to look through a high-powered telescope at another entire galaxy full of a hundred billion stars like our own Milky Way. But many more people have been to the beach. They have sat on the warm sand and dug their feet in deeply, letting the grains cover them up to their ankles. They've stared down at the sand covering their feet and tried to separate it into individual crystals, each one catching the light and reflecting it into the edge of a nearby grain, or up to the sky, or to their eyes.

We say "observable universe" because the universe is expanding. Since that first tiny instant when the universe exploded from nothingness into being, it has been moving outward in all directions. All that we can see at the present time are the objects in the universe whose light has had time to reach our eyes here on Earth. But that light will change as time goes on and matter spreads out farther and farther in space. There is still more to see that we haven't yet.

I knew something about this from my life. I reread one of my journal entries recently. There was a day when I had fled to the beach when it had all felt like too much. I don't have any memory now of what it was that felt overwhelming, but it was probably everything all at once. I started out as a student, not knowing anything. As I progressed, I learned more. Then people expected me to know even more. And then there is a fear that moves in sometime during that transition and makes a home. It sets up a table and chairs and puts up its feet. This fear was not invited, but it doesn't really care about details. It's there to stay, unless some truly powerful juju magic is applied to drive it out. What I needed was an exorcism. Either because my field was so narrow and new, or because I was one of the magical unicorns, the rare Dúnedain in Tolkien's literature—African American, female, and a scientist—the fear was so magnificently gargantuan. The stakes felt so high. *What if I don't know anything, really? I've memorized a bunch of things, and I am just regurgitating them. And if they ask me about x, y, z, I'm hosed.* Many minutes leading to hours of looking up x, y, and z followed in the hotel room the night before a talk, during which I tried to consider every possible question, and its derivative follow-up question, so that I could have an answer ready. The memory of the deer-in-the-headlights experience of my first journal club talk back in grad school was suddenly not five years old but fresh as the rain I watched fall as I wrote this. It is a kind of trauma, and it has its own sense memory, the way we learned how to resurrect the feeling of being drunk, or hot, or cold, in acting school. We were taught to key in to sensations in our memory of those states—the small beads of sweat that form under the nose just above the upper lip, the desire to walk straight, the trying, the sheer magnitude of effort with which a drunk person desperately tries to walk straight, and in the failure of that effort, reveals the extent of their drunkenness. We never *tried* to stumble. We tried to stand up and walk. The incongruence between the intent and the result made the effect truly believable.

In that journal entry, I had written this:

On the precipice of some great unfolding, a transition into a new breadth, a new ocean of possibilities and change. I want to let myself unfurl, like a great ancient sail or enormous bird's wing. I want to blossom before my own eyes.

In February, I left Los Angeles for a multicity trip. I was scheduled to give two seminar talks—one at Yale and one at Harvard—and a short lunch talk at Harvard about Rising Stargirls. What I was most excited about was the trip I would take before these talks. I went to Taos, New Mexico, to see my writing teacher, Natalie, and celebrate the thirtieth anniversary of her first book, *Writing Down the Bones*. I wanted to be there. That book and Natalie's workshops over the past fourteen years had meant so much to me, and to my path as a writer, a scientist, and indeed, a human being.

I flew to Albuquerque, rented a car, and drove to Taos to meet Natalie for chocolate at Kakawa. I sipped warm, chili-spiced chocolate mixed with tamer, rich, "American" chocolate, and we talked like old friends. We were becoming that, after many years of my putting her on a pedestal. Finally, now forty years old and paying my own way, I could see her as a human being, a person who'd made choices, including what path to follow. She had been writing about that path for the last thirty years. Here, now, she could see all that meant: we keep going, she said. Another thirty years.

I stayed at Mabel's again—a historic house built by an Indigenous American man from the Taos Pueblo and an upscale New York socialite who'd come here and never left. I stayed in the top room, a room surrounded on all sides by windows. I watched the Sun set in a dozen brilliant shades of flame—orange, yellow, bursting out of the horizon. I woke up to the few rays peeking out from behind the hills to the east. I saw the sunrise before I saw the Sun itself. I sat outside and listened to the hundred species of birds in trees, to the most distant sound I could hear—the car zooming down the Paseo in town. I heard a dog down the road chasing squirrels, the slamming door of a house across the parking lot, the chime of bells in the square signaling—was it 10:30 a.m.? How long had I been out here? It didn't matter. For once, I had nowhere to be.

A crow was calling. Or maybe it was a magpie. I could look it up. But why? There was nothing I had to know or tell anyone here. I wasn't at my talk venue where I had to know things. Someone turned on an electric saw. *They're building something close by*, I thought. Maybe sanding a wall meant for someone's new house. That's nice. The wind was crisp, yet warmed by the sun on my face. Spring had come early to Taos. It blew the leaves, crackling and dry along the ground, and the wisps of hair on my cheeks. The magpie's call was lower. *She sounds tired*, I thought, *or like she has given up trying to get somebody's attention. That's, of course, when he'll come.*

I once had a T-shirt that I wore until it was in shreds around my shoulders. It had a Southwestern landscape with pink cliffs and rust-colored mesas, a wolf with its head turned up to the Moon, and a quote that read, IN THE SOUNDS OF THE WILD, WE AWAKEN TO WHO WE ARE. Amid the sounds of the world—everything, natural and man-made—I remembered who I was without the din, glory, or even courage. I realized that I didn't even have to have that to be me. Inside me a new thread of knowledge was forming. It was an awareness that I mattered, regardless of what I did or said, and whether any of it was right. It was taking its time to string its way from my heart to my mind. But it was close enough in my consciousness that I felt its growing presence. I trusted that when it arrived fully, it would be woven in deeply, to help me remember. I needed to believe that.

I just heard a bird's wing flutter. I'd never listened for that at home. But I bet if I did, I would hear it.

Evolution

When I became a TED Fellow, I received more than just the opportunity to tell the world something that I believed was important. I also received access to a life coach for twelve months. Every week we talked on the phone for an hour about whatever I wanted to—my career aspirations, personal life issues, or how to manage my newfound visibility after my TED Talk had gone live. It seemed I had gotten on everyone's radar. People I didn't know were reaching out from across the world. Mostly they wanted to say how much they'd appreciated what I'd shared in my talk. Sometimes they wanted encouragement and support as they embarked on a new career that they too felt had broken some mold or veered decidedly away from the norm. It was incredible to learn, after thinking for so long that I was the only one whose dreams crossed seemingly disparate disciplines, that I was nowhere near alone.

But it was also overwhelming, especially when the emails included requests that involved time commitments on my part, on top of what I was already doing. It had dawned on me that if I kept saying yes—to this invited colloquium talk, and that public lecture, to this interview for some TV science program, and that twenty-minute phone interview for some magazine piece—all those yeses were soon going to add up to a significant amount of time that would not be available to me to put into doing the things that were actually going to get me a permanent position in academia—publishing and getting grants. My coach, Terrie, gently and frequently suggested hiring an assistant. Eventually I agreed.

I hired an undergrad at UCLA as my assistant for five hours a week, largely to help me field those endless requests. Often she just responded with a polite no on my behalf, parroting a list of phrases I'd recite to her. Having my assistant as a buffer between me and the rest of the world was a relief, and it spared me the full weight of the emotional cost of saying no, which seemed to sit heavier in my belly than it did in other people's.

As the lengthy faculty job application process came closer and closer to actually materializing into a job offer, I began to share my fears about whether this was what I wanted, and whether this was who I was—a professor. Maybe I didn't need this life of long hours and pressure to either continue to produce cutting-edge research or be unceremoniously put out on the street. As Terrie listened to me ramble on about how I had always been a nontraditional person, and maybe that meant I could never be satisfied doing a traditional job, she finally said something that woke my brain up in its cold sweat.

"So you're not sure if being a professor is for you. Well, there is only one way to find out."

Maybe I didn't have to know what I was going to do for the rest of my life. Maybe all I had to do was decide what I wanted to try doing right now. If I didn't like it, I could leave. It could be that simple. I had evolved, like one of the large multicellular organisms that had developed on this planet, gulping fresh oxygen that had never before been abundant. My small, single-celled life dictated by the qualifications necessary to have a future in academia was no longer sufficient for my happiness. I wanted it all. I was not afraid to be who I was anymore. I was ready to embrace it.

I accepted a job offer as a professor of physics and astronomy at UC Irvine. Even stepping just one foot on this path, there were still many questions that remained. Could I continue on my evolutionary path of merging science, writing, and public speaking, and be a professor at the same time? Or would I need to devolve and conform to get tenure—that coveted, untouchable state of permanent job security unique to institutions of higher learning? I would be the only woman of color in my entire department. What would be the emotional cost of that reality? How many times would I have to let go of being mistaken for an administrative assistant? How much thicker would my skin need to become to handle receiving poorer evaluations from students solely because of the color of my skin? Even if none of those things happened, would I even like being a professor? Virtually every faculty member I had interacted with as a student seemed overworked and stressed out. But there were a precious few I'd talked to who appeared to have accomplished the unimaginable—working the nine-to-five shift reserved for office staff and construction workers and still getting tenure. They walked among the masses as if ethereal. What had they done? I had to find out.

I enrolled in professional development programs that taught me how to prioritize research and writing, then teaching, and lastly, service. Service was what the administrative leaders of the department and the larger school, and even other faculty members, wanted you to do: it was a way of showing that you were part of the department and the campus, that you cared about its future enough to serve on committees to help it run smoothly and prosper. And it took up a lot of time. But when it came time for you to get promoted, what mattered was how many papers your team had published and how much money you'd brought in as grants to the university. At the same time, if you did no service, you weren't seen as a team player within your department or the larger campus. Those who could play the game strategically, knowing when to say no and when to say yes to service that served their own best interests, while guarding their research and writing time like a sphinx, prevailed. Like any game, you could get better the more you practiced. By the time I showed up on campus as a professor, I vowed that I would be a whiz at it.

☽

Whenever I go on vacation, something big happens in my field. Like, "more-planets-found" big.

When I went on vacation in August of 2016, after my summer stint at Harvard, the CarMax salesman told me about the planet found orbiting Proxima Centauri. I sat at his sales desk looking up press releases while Steven test drove the 2013 Hyundai Accent we were there to buy. It was going to be Steve's car, so I let him take the wheel in all respects. I sat there feeling out of touch with my field, half interested in the discovery and half obligated to learn at least a few basic facts about this planet in case I was asked about it by the grocery store clerk later that day. I mostly felt guilty for not being dedicated enough to my field to stay plugged in while I was on vacation.

Becoming a professor was, honestly, something that sort of fell into my lap. When I thought back to the original reason I'd decided to return to graduate school in astronomy, it wasn't because I wanted to become a professor. I wanted to know more about astronomy. And if I wanted to be seen as an expert, whether as a science TV host or as an astronaut candidate, I needed a PhD. All roads had led to going back to graduate school.

Getting that actual doctorate had seemed like such an insurmountable task that, planner though I am, I hadn't given much thought to what I was going to do with my PhD once I got it. Apparently I was going to try my hand at being a professor. But the science choices, research publication expectations, and requests for my time all conspired to make me want to crawl under a rock. I missed the days of being a grad student, when no one knew or cared who I was, nor did anyone expect me to do anything except learn, ask questions of people more senior than me, and eventually, somehow, graduate.

Now I felt put upon. And I didn't like feeling that way. It reminded me of when my mother would fantasize out loud to us about how she dreamed of being committed to a hospital so she could rest, and no one would expect anything from her. I began to understand how she had felt when she was a professor. That scared me.

I talked about this to any trusted person who'd listen. Not my husband, because he'd worry. Finally, I escaped to Ojai, for the first of two writing retreats I'd arranged for myself before starting my faculty position. One, later in May, would be organized by someone else. This one was organized by me, for me and me alone. Retreats have become a critical component of my life, especially as an academic in the sciences and a woman of color. They help me to reset, recharge, and get some perspective on my life and all the things in it that feel like big deals, and as I typically discover over the course of these retreats, usually aren't.

I can still remember the sound of my feet on the footholds and rocky crevices of the mountains of the Los Padres National Forest, on my hike of the Pratt Trail on the last full day of my retreat. The crisp, bright air and sunshine, the smell of fresh earth still wringing itself out from a week of rain, the evergreen mossy smell of the tree canopy towering overhead. Here, there was no doubt about the presence of life, and how much life was here. How little that life cared about my feelings of insecurity in my profession or my promise as a scientist. Here, animals breathed and flew and scurried and dove for food and survived. They simply did the next indicated thing, while the sunshine and the clouds and the rain swirled overhead and sometimes stopped and stayed a while, and then eventually moved on. I had learned years ago that when in doubt, I could do the next indicated thing. It was much easier to do, or so it seemed, when no one else I knew was around to judge that thing.

I knew an announcement was coming. I'd been checking my email the first two days of my retreat and had received an email from a reporter at the journal *Nature*. She was looking for other scientists to comment on a "big discovery" that was going to be published in two days and had read my review paper on the habitability of planets orbiting M-dwarf stars. She was hoping to speak with me the day she sent the email (of course), or at the latest, the next day. She was willing to send me the embargoed pdf of the paper if I was interested in talking with her. I left the email in my "emails pending" folder and walked over to the breakfast buffet table for more cheese.

I never responded. She eventually got one of my coauthors to comment on the discovery. There was a one-liner from her in the article regaling its far-reaching implications. Good.

When I finally read the article and thought about how much time I would have spent reading that discovery paper and prepping for that interview with that reporter, and how all of those hours would have been distilled down into one single sentence, I laughed out loud. I realized that truly, some things are just nowhere near as important as others. I would have had to trade a good chunk of my retreat time just for the privilege of knowing about the discovery a couple of days before the rest of the world. I used to lament being one of the masses learning about scientific discoveries on the news along with everyone else. I'd left the field of astronomy and was no longer in the club. Now I sometimes chose not to change my plans and add extra time to what was already a busy life, just to be in the know.

The discovery was of the largest number of Earth-sized planets discovered in one system thus far. A total of seven, all orbiting the same star—an ultra-cool M-dwarf star named TRAPPIST-1, after the ground-based telescope that found the first two planets orbiting this star. Now there were five more. And these planets were a mere forty light-years away. Certainly not reachable in our lifetimes, but compared with the planet I'd recently studied—Kepler-62f—these planets were close. Not next door like Proxima Centauri b, but certainly down the block.

I decided I would check the arXiv—the website where astronomers post papers in advance of publication—just to see what all the hubbub was about, as all I'd gotten from the reporter was that the discovery involved a planet orbiting an M-dwarf star. I knew it had to be more than that. NASA was holding a press conference in the morning, for Christ's sake. I mean, any planet discovery is a big deal, of course. Twenty-two years ago we knew of no planets orbiting other stars like our Sun, or any other hydrogen-burning stars, for that matter. But now we knew of close to five thousand. The ante had been upped. Headlines needed to be bigger, brasher, sexier.

The next morning I forgot about the whole thing. I was in Ojai. I'd finished the only scheduled meeting I'd committed to on this trip, via Skype—an interview of a faculty candidate in observational exoplanets who, comfortingly, also had no idea about this pending announcement. Now it was just me, whatever I chose to work on for the next two days, and the sunshine, which had finally emerged from the Seattle-like clouds, rain, and gloom that had blanketed Southern California for more than a week.

My website homepages loaded as I sat in the breakfast room eating homemade granola and yogurt with a side of smoked gouda cheese. The headline on CNN read, "Seven Chances for Life." I wanted to know more. *Those damn headlines. These journalists are good.* I spent the next hour tuning in to the NASA press conference replay, reloading the page when the voices and their attached faces froze. The bandwidth on the Wi-Fi was nonexistent. I hung in there. Three of the seven planets were in the habitable zone, which meant they could have liquid water on their surfaces. All seven of the planets should be able to be followed up on with James Webb, the next big space telescope and Hubble's successor. All of NASA's main telescopes—Hubble, Spitzer, Kepler (in its adapted K2 configuration)—were trained on them now. More information would be coming. And more planets would, no doubt, not be far behind in the race to be revealed to the world as the best potential prospects for life.

While I listened to the press conference, I paid attention—not just to the astronomers, all of whom I'd either met or corresponded with, but to myself. I noticed how I felt as I learned more and more about the TRAPPIST-1 system—future plans, current impressions, speculative assessments, and implications for the field of exoplanet discovery and characterization.

The old familiar feeling of overwhelm began to engulf me, creeping up my arm like the frigid chill of early morning lake water as you ease your body in, hoping it won't jar you too much and stop your heart cold if you go slowly. But it still shakes you awake and sober. I was used to feeling overwhelmed when it came to science. All of the expectations—internal and external, perceived and legitimate—had felt, day in and day out, like a coat I was just going to have to get used to putting on every time I sat at my desk. Whenever I opened my email the coat became a parka, with a 20-pound dumbbell in each pocket. I didn't know how I was going to manage it all, and I had begun to question whether I even wanted to.

I've come to realize that as an African American woman in the sciences, I will always be asked to do things, give my time in various ways. The weekly telecon I'd kept up for years with three other women of color in astronomy confirmed that I was not alone when it came to getting lots of requests. I don't have to say yes, and most of the time I say no, and will most likely continue to do so, for my own sanity, serenity, and physical health. I also don't have to be upset, or even mildly disturbed, when I'm asked. My equating someone's request with some sort of additional, significant, time-consuming action on my part is what sets me spinning and cursing and in danger of dreaming of being a hermit living on the side of a mountain somewhere in the back country, having left science again and once and for all, because it just demands too much—too much time, too much effort, too much worry, too much more of my life than I want to give. I don't have to live in extremes.

When I listened to the press conference, I discovered that my familiar feeling of overwhelm didn't come from a disinterest in science and pressure to do it. It came from fear. Huge fear. There was a fear of being part of those future press conferences, as I know I will be. There was a fear of saying something that's not quite right, or just downright wrong, and though imperceptible to the general public, sounding stupid and incompetent to my peers for getting some basic fact wrong that anyone in their first year of graduate school should know. There was a fear of wanting so badly to work on this new planetary system—to write a paper of my own on one or more of these planets—and getting so excited to do just that, only to see my hopes dashed after seeing someone else's paper, on exactly the same project, on the arXiv. There was *what if* after *what if*. They crashed against my mind and body like waves against the shore. They lodged in my gut and sat there, gurgling.

Then, the next thought: *Oh! This is what this is all about. This is why it's scary to be all in, and much easier to be one foot in, with the other foot on a fantasy roller coaster headed back for Hollywood. Or to dream up profession #2—the life of a writer, where my time is my own, I answer to no one, and I don't have to interact with another human being at all, other than my husband, if I don't want to. Thank God for online shopping.*

I got to stand outside my fear for the first time in a long time—maybe ever—and recognize what was underneath all the papers strewn all over the floor—a very nice, deep cherry hardwood floor! And that floor was varnished with excitement. I was afraid to hope for things and be disappointed, or to get things wrong and be embarrassed, because I care about these things. I'm excited about them! I just might want to work on this new planetary system. And history tells me there is always more work that can be done on something, no matter who publishes what. I didn't know it then, but my first postdoc would publish a paper on the effect of land surface composition and albedo on the possible climates of these TRAPPIST-1 planets—the first of its kind. It is an abundant field. The universe is teeming with planets waiting to be explored.

Pregnant

As I waited, I stood there looking at the box through the glass, still slightly dubious after the initial test that morning. Here I was, forty-one years old and buying a pregnancy test in a drugstore for the first time in my life. The other one I'd ordered on Amazon. But I needed a clear indicator, not three lines of questionable shading that I was supposed to interpret. I wanted a plus or a minus, a thumbs-up or thumbs-down.

With my daughter cat Mimi at my heels, I flitted around the house throwing on clothes. I got in the car and drove down the street to Walgreens. I found the aisle for feminine hygiene products and saw it—the test that would tell me by way of a single word whether the first test had been accurate. I would see either PREGNANT or NOT PREGNANT.

The tests were locked behind glass and accessible only with a key, so I signaled one of the drugstore attendants, and she walked to the back of the store and disappeared through a door in the corner.

A comedy of errors ensued. The attendant returned after what seemed like fifteen minutes, only to discover that she had the wrong key after trying to jam it in the lock. She disappeared again. I started to wonder if I would ever emerge from the store with the test. Finally, the glass case was unlocked, and like a greasy-fingered child at a carnival, I was allowed to reach inside and claim my prize.

I sped home, ran upstairs, opened the packaging, took out a test stick, pulled down my pants, and peed on the stick. I waited the requisite three minutes, then peered over at it with one eye closed, almost afraid of what I would see. The single word in the display box was definite: PREGNANT.

"Holy shit," I said out loud, probably not for the first time that morning. "Is this for real?"

All I could think of was telling Steve. How would I do it? When? He was at work. I had to meet people in half an hour. I remember almost nothing about that meeting. I was so focused on the moment when I could drive over to Steve's workplace and tell him the news. I was full of secret delight. Nothing could touch me.

I hadn't lost sight of my initial reaction to the news myself. I made a point of noticing, through every stage that morning. After the first test there was . . . hope, excitement. After the second, there was . . . incredulity, a little bit of disbelief, and . . . joy. I hadn't been completely sure how I would feel either way. This was why it had been so confusing for so long. But these feelings were clear and undeniable. They wouldn't be the only feelings, to be sure. But they were here now, and they were real.

Steve and I went to Tender Greens for lunch. He plopped our tray of plates down—roasted potatoes, grilled salmon, and kale parmesan salad for me. Steak, mashed potatoes, and seasonal vegetables for him. He sat down across from me in the booth and started eating. I watched him, my belly fluttering, anxious and unsure of how to say it.

I blurted it out. "So I took two pregnancy tests this morning, and they both came back positive."

He stopped mid-forkful and looked up at me. His eyes opened wide.

He sat back hard against the back of the booth and stared, incredulous. Then he opened his mouth.

"No shit."

His voice had laughter and giddiness in it. Then his rational mind came back.

"What are we gonna do?!"

Laughter. We sat staring at each other. He reached across the table to take my hand and rubbed the top of it gently with his thumb.

"I want to name him or her after my father, Garland."

That was quick. We were already on to baby names. We'd lost Steve's dad while we were in Seattle. I knew how important this was to him. I hadn't even thought about what names I might like, but it wasn't the time to argue. I said, "Okay."

After lunch, Steve was headed back to work. As he helped me into my car to head home, he kissed me and said, "I'm so proud."

Weeks later I found myself at a symposium at Stanford to give a talk on the M-dwarf planet habitability review paper I'd written with John Johnson and Sarah Ballard. The venue was the Breakthrough Discuss Symposium—a gathering of scientists, engineers, and high-powered software giants in Silicon Valley, all brought together in one place to discuss the best ways to figure out if planets around the small, cool stars were habitable, preferably inhabited, both from the ground and from space. Breakthrough Discuss was an offshoot of Breakthrough Starshot—the developing mission and brainchild of the likes of Stephen Hawking, Google cofounder Yuri Milner, and astronomers brought into the project to actually assess the feasibility of launching tiny spacecraft at 20 percent of the speed of light, using lasers propelled toward the star systems nearest to us.

In short, the symposium was a big deal.

PREGNANT

This was the first conference I'd been to since finding out that I was pregnant. Since it was in Palo Alto, California, I decided to tag onto it a visit with my dad in nearby San Francisco when it was over.

I remember wondering if people could tell that I was pregnant, even at ten weeks. I remember that I tripped over my words, as I do sometimes when my brain gets ahead of my mouth, which is embarrassing to admit as a classically trained actor. It was only for a few seconds, and then I returned to clearly articulated speech. I remember that I wore a killer outfit—pink fitted slacks with a crocheted tank top with layers of gold, rose, and black that matched them perfectly. I remember meeting Jill Tarter—former director of the Search for Extraterrestrial Intelligence Institute (SETI), and the inspiration behind Jodie Foster's character in *Contact*—in the bathroom. She was nice, and she said something that I don't remember that made me feel welcome in the field. I remember the huge party at Yuri's estate at the end of the symposium that we were bused to. I remember that hors d'oeuvres and desserts were everywhere, and how the drinks flowed everywhere but down my throat, and how I stood next to my PhD advisor, Vikki Meadows, also there for the symposium, and felt almost like a peer. I remember that it wasn't long before I was ready to go home and took a golf cart back to the bus with one of my old planetary science professors, though the party would continue long into the night. The air was cool and I was relaxed. My talk was over, and scientists who were drunk cared a lot less about how good they thought my science was or wasn't. It didn't matter anyway. Something was growing inside me that only wanted to be loved.

Welcome to the World

There is so much to feel when you're pregnant. It didn't matter who made some annoying comment at work, or what the state of my career was. My own private joy enfolded me, protecting me from it all. The immediate goal was no longer to identify other planets in the universe that might host life. I was now hosting life! That life was growing bigger every day. What mattered more than everything else was keeping myself healthy so it would stay healthy and continue to grow.

I had hired a new coach, Mary, who specialized in academics. I had found her through the National Center for Faculty Development and Diversity (NCFDD) not long after I had secured my faculty position. I knew that as an African American professor—in the physical sciences, no less—I would continue to receive more requests for my time, many more than my white male and even white female counterparts. And I hadn't the foggiest idea how to determine which service requests were ones I could comfortably say no to, and which ones were good strategic ones to say yes to. I didn't know how to successfully navigate those departmental relationships with colleagues who would one day be voting on my tenure case. The entire landscape of my department, where I was the only Black

WELCOME TO THE WORLD

faculty member among fifty-plus professors across all branches of physics, seemed like a sociopolitical minefield. I needed help.

Mary and I talked about everything. My career, the move to Irvine, my husband, our decision to finally stop preventing pregnancy, difficult conversations with coauthors about the direction papers were going, how to choose which grants to apply for. NCFDD had taught me the value and benefit of writing every day for at least thirty minutes. In one year I had seen my productivity nearly triple, with multiple first-author papers submitted that year. I loved writing, so applying that principle to daily academic writing wasn't as much of a stretch for me as it was for other professors. But it had still been a discipline to prioritize it, schedule it like any other meeting or appointment, and make sure that I'd written for at least half an hour before I even thought about checking my email. I made my strategic plans for every quarter and mapped out the projects and tasks associated with completing a manuscript, leading up to submitting it to a peer-reviewed journal.

Mary and I had strategized when I got pregnant. I made a special trip to see my future department chair and other colleagues who'd helped recruit me to UCI, to give them the news. It was a diplomatic mission. I didn't want anyone thinking that I had secured a job for the health benefits, without any thought to leaving them short an active member of the department for nearly an entire academic year while I was on maternity leave and modified duties. Rather, as Mary and I discussed, I assured them that although we'd been hopeful for a while, this was all a pretty big surprise now, especially given my age, and that we were thrilled. The "being thrilled" part was important too. This wasn't an apology for being pregnant. It was unfortunate that this kind of delicate touch was necessary, and maybe it would have all been fine if I hadn't applied it. But I was glad that I had. It set a tone of respect. Words are powerful, and so is framing.

The move to Irvine was intense. Steven wouldn't let me lift a finger. We were on a priority list for a new house, and our number had been called, but the house was still being built and wouldn't be ready for six months. So we lived in a two-bedroom apartment that was a ten-minute walk from campus, and kept as much as we could in storage. It looked like our house was going to be ready right around the time I was expected to give birth. That was going to be interesting.

During my pregnancy, a few weeks after starting at Irvine, I submitted a proposal for the NSF CAREER award, a five-year grant for early career faculty. Mary and I decided that I would teach right away rather than use my teaching relief first, as so many faculty do to ease into academic life. Once I had the baby I would be on maternity leave and modified duties (no teaching) for the rest of the year. It was a strategic decision to teach while pregnant, to show that I was a team player and contributing from the start, even while carrying around twenty-five extra pounds. I taught right up until I went into labor. I felt great. Daily walking to and from work and prenatal yoga had contributed to that, and to offsetting any potential dangers of a forty-two-year-old "geriatric" pregnancy.

The course I taught that fall of my first year as a professor was Life in the Universe. On the first day of class, my husband had a pot of roses delivered to my office that read, "Congrats, Professor Baby!" It warmed my heart.

Dancing the edge alongside my sheer delight at being pregnant, though, was absolute terror. What if something went wrong? I could nurture and grow this baby inside me for the full nine months, and it could still all go to shit at the end. It had happened to my mom. And her mom before that. Stillbirth. I suddenly, sharply understood the next-level scale of pain that could be. All the hope for this being, the plans and dreams, the family meals and trips and baby clothes and bedtime stories and giggles and laughter and cuddling and seeing our entire family in their eyes—

shattered. Erased in an instant. I couldn't fathom. That phone call from my grandmother telling me about Mom's stillbirth when I was at MIT became something altogether different in my memory.

I listened as my mother warned me, sharing her experience of that pregnancy. And then I couldn't listen anymore. It was my pregnancy. No one else's. I couldn't dwell in a dangerous world of *what if* . . . I busied myself. Whatever the doctor said to do, I did it. Drink a gallon of some yellow fluid and don't go to the bathroom until I was poised to explode, so they could ultrasound the entire baby's body? Sure. Drink another fluid, this one syrupy sweet, and get my blood drawn to make sure I don't have gestational diabetes? Okay. Run the chromosome test for Down syndrome (at age forty-one, my potential to have a baby with Down syndrome increased by factors of ten, compared to young twenty-somethings or even thirty-somethings with child)? Do it. Count, at least three times a day, the number of times I felt the baby kick me, and write it down in a notebook, and make sure I could count at least ten kicks each time, and if not, call the doctor? Yep. They were asking a vigilant person to give her vigilance a steroid shot for the next six months. I could do that. I was born to do that.

I came in three times a week so the fetus could be monitored, my blood pressure taken, sensors attached to my belly to measure its heart rate, charting it with reams of paper spitting out from the computer. I had just started a faculty position and moved from another city, half my stuff was in storage, and I had no idea when we'd be moving again. I knew I should be writing that new paper on the climatic effect of ice that is so full of salt that it forms a crust on its surface that's even brighter than snow in the infrared. And what does that do to the habitability of planets orbiting M-dwarf stars if we include this kind of ice in climate models? My glaciologist colleague Regina Carns and I wanted to find out. It looked like planets that were in the habitable zones of these stars could actually get cold enough for this type of salt-crusted ice, called hydrohalite, to form, meaning these planets could be colder than previously expected, which could be a big freaking deal. And we had to get this paper submitted before I gave birth. Once I gave birth, I wouldn't have time to do anything for months. I would be taking care of another human who wouldn't be able to do anything for itself. It didn't feel like I had any amount of time to sit in a room on a leather recliner that stuck to my legs (I was living in maternity dresses from Seraphine; I would be a fashionable pregnant professor). But I did it anyway. I listened for the regular rhythm of my baby's heartbeat, not too fast or too slow, and prayed that it stayed just like that for every test, every time I went there every week, and that my blood pressure wouldn't go through the roof if it didn't, sending off a red flag for some other tests to be done.

When the genetic specialist called me and told me that the chances of my baby having Down syndrome were one in ten thousand—a probability that, while not nil, was the very lowest that could be calculated in this case, offering the greatest chance possible that the baby wouldn't have this syndrome, the relief was like ice melting off me. As I lay in bed, I felt all the tension seep into the covers, leaving only me, pure and unburdened of worry for that single perfect moment. She asked if I wanted to know the sex. I said yes. In her melodic, East Indian accent, she replied, "You are having a beautiful princess."

I called Steven at work. He was at the booth, busy selling tickets before the matinee, but he got someone to take over for him, anxious to hear the results. When I told him the good news, I felt his sigh through the phone. Finding out that the baby was no longer an "it" but a "she" was new and thrilling. We would get to that level of excitement soon enough. Finding out that she was healthy was the space we rested in for a good long while. We let it fill us.

WELCOME TO THE WORLD

Steven and I headed back to Ojai for my birthday. I wanted to introduce her—our daughter—to our favorite vacation place on Earth so that we could rest in the quiet solitude of the retreat with her. We listened to the branches move, watched the bright sunlight hit the rocks, bringing small lizards out of hiding to do push-ups in the heat. I napped, slow walked the grounds, painted, and wrote. Steven golfed and then came back to our suite to join me for lunch on our patio overlooking the valley. We nibbled cheeses, fruits, and decadent cinnamon-topped dessert pizza delivered from the neighborhood place in town. I panicked for about a day when I felt a strange ache low in my belly, and my mind went wild. The next day it was gone, probably some residual strain from sitting in the car for two and a half hours. I marveled at how something can ride the wave of speculation into forever if I'm not careful.

We'd vowed that our daughter would love books, so we started early. We read her *Goodnight Moon* before bed each night, imagining her falling asleep in my womb to the sound of our voices. We watched my belly move up and down as I breathed in the lit afternoon, lazing away the day in bed. The night before we left, we visited our favorite formal restaurant in Ojai, the Ranch House. It was an outdoor restaurant with a lush garden in the backyard that guests could wander through. We lingered over our five-course meal, wondering what it would be like to bring our daughter here as a little girl. Would she like it? Would she throw crumbs into the koi pond? Would she stick her face full into the roses and the other fragrant flowers, the way her mom does, just to breathe in the sweetness? We tried not to get ahead of ourselves. But it was hard.

In my last trimester, Steven took me to Laguna Beach, a short twenty-minute drive from our house in Irvine. This was our "babymoon"—the last precious time to be alone together on a trip before we became three, with all the chaos that would follow. Even though forty weeks is the plan for a pregnancy, a baby can be expected to make an appearance at any point after thirty-seven weeks and still be considered full term. Babies don't always wait even that long, of course. But our daughter had made it through being physically turned upside down through my skin by my doctor once it was discovered that she was still in the breech position at thirty-five weeks. That was an experience—my doctor pushing and twisting my belly like he was trying to turn a stiff pipe nozzle to open. It had worked. She hadn't decided to abandon ship and head for an exit door after that, so she was probably going to stick around for the duration. Still, it was a good idea to stay close to home.

Steven found a hotel overlooking the ocean. The room could have been in a Motel 6; we were paying for the view and the sound of crashing waves that didn't come from a sound machine. We ate on the deck looking out at the Sun sinking into the ocean. We strolled through the narrow cobblestone streets, stopping in shops to gaze at odds and ends. We discovered gelato nearly as good as we'd eaten in Florence almost a decade earlier, and slurped up the smooth, cold creaminess before losing a drop from the waffle cone that cradled it.

At one store I found it—the souvenir I'd been looking for. A single delicate silver chain, feathery light, with a circular wafer-thin pendant and a blue-green stone hanging in front. On the pendant three words were engraved in script, one below the other.

> *Thankful*
> *Grateful*
> *Blessed*

Indeed.

WELCOME TO THE WORLD

☾

During week seven of the quarter, Mary and I had a phone call. I was at home working, and feeling light contractions. She asked me if I'd finished writing the final for the class I was teaching. I said not quite. She suggested that I work on it during the day, so I did.

By the early evening the contractions had gotten regular, yet were still far apart. I called the labor and delivery number I'd had on my phone since the beginning of my third trimester. A woman with a far-off voice that sounded like crystal glasses tinkling together told me to call back when my contractions were less than one minute apart.

Steve came home. We watched *The Great British Bake Off.* I sat on my big, green exercise ball, the one I used to practice the labor exercises that I had learned in our Preparation for Childbirth class at Kaiser Permanente Orange County–Irvine Medical Center. Steve kept track on his phone of the length of and intervals between my contractions. Nine minutes between. Then four. I waddled back to the bathroom, and in the hallway called back to him, "Here's another one!" He started the clock again. Then twenty seconds later it was done. "Okay, it's over!" I continued back to the bathroom. Less than a minute later I called, "Another one!" This one felt stronger and more pressing, like someone trying to nudge me awake in the dead of a deep sleep, deep like the deepest oceans, and I wanted to stay down there, it was so peaceful: *This is the deepest sleep I've ever had and I wish I could stay here forever, but this person really wants me to wake up—the nerve.* It was clear and certain, in the wake of that pulling and squeezing compression—this was the night. She was coming. I came out and said one firm sentence to Steven, and he knew it too: "Put your shoes on."

Getting to the car was its own adventure. While we waited for the house we'd been offered as part of my faculty start-up package to be built, we lived on the third floor of an apartment building that had no elevator. It had seemed romantic at the time. We would take walks up the hill to peer through the construction fencing and watch our home change before our eyes, from a skeleton, to a shell, to a home with paint and windows that would be ours. I would own a home before either of my parents. Whoa.

In this moment, though, I was not feeling very romantic about where we lived or what was in store. I just wanted to get to the hospital as fast as humanly possible. At the end of each half-flight of stairs, I had to stop to breathe through another contraction. When we were finally in the car, Steven drove like an old person. He had listened in childbirth class. *All of the excitement, the urgency, will make you want to race. But take it slow and easy. You need to get there in one piece.* Totally unlike the wild car rides to the hospital I'd seen in the movies. I told him to drive faster, but he kept right on going at senior citizen pace, eyes never leaving the road, right hand holding mine during contractions when he didn't need it. Good man.

We got to the hospital about fifteen minutes later, around 1 a.m. Our daughter wasn't born until 6 p.m. that day. My friend Susan came and gave support along with Steve, helping me breathe, staying with me when he needed a break to eat and get some air. Then they switched off. They were my labor team, along with three shifts of nurses and doctors who got me through an imbalanced and then readministered epidural, the roller coaster of our baby's normal then plummeting heartbeat with every contraction, the Pitocin to speed up delivery, followed by the terbutaline to slow it down when the heartbeat was plummeting into the danger zone. There had been forty-five minutes of pushing when I finally made it to 10 centimeters, only to realize that her head hadn't dropped by any amount—she was still at 0 station. And then came the final decision to do a C-section to get her out. As they told us in childbirth class, you can prepare as best you can for your ideal birth experience, but in the end you have to build in the flexibility to roll with whatever happens and make adjustments. My obstetrician happened to be on duty in the hospital at the time. He would be able to do the C-section. It wasn't how things normally went with this medical organization. It was a gift. Steven stared down at me in the hospital bed and said he didn't care how we got there. A healthy baby was all we wanted.

The anesthesiologist said he'd never seen someone shake as much as I did. I couldn't stop. Whatever they gave me, combined with the hard cold of the operating room, left me with no way to calm it down from the chest up. I was glad the rest of my body was still enough for them to open up my stomach. When they took her out they discovered the reason behind her yo-yoing heart rate during labor: my umbilical cord was wrapped around her neck twice. I don't know how things would have gone if I had tried to keep pushing her out. When she came out, she was perfect.

☾

I sent my last work emails from the hospital bed, as our baby slept in her bassinet beside me, Steven on the couch nearby. To the 130-plus students in my Life in the Universe class, I wrote, "If you're reading this, it means that it's time for me to contribute to the known complement of life in the universe. I wish I could have stayed with you through the end of the quarter. But Baby Shields had other plans."

To the trio of colleagues who would share the responsibility of finishing out the last three weeks of the course for me, I wrote letting them know that we'd had our baby that day, November 16, 2017, at 6:11 p.m., that she was 6 pounds, 7 ounces, and that her name was Garland-Rose Delphine Shields (we had named her after Steve's father, my favorite flower to smell, and my grandmother, the one who loved numbers, who had died a couple of years before Steve's dad). I attached a picture of her sleeping peacefully in her bassinet, exhausted from her journey across space and time to here, now, and us. I told them we were all doing fine. I thanked them for taking over the course for me, and told them where to find the PowerPoint slides for the rest of the lectures and the draft of the final. I told them about the colleague from UC Riverside whom I'd lined up to give a guest lecture on December 5. I knew they'd like that; less work for them to do. I told them they could edit anything I'd put together as they saw fit. And I pressed SEND.

Then I enabled the automatic maternity leave email message I'd had ready for weeks, and signed out of my email account for the next eight weeks.

My dad was on his way from San Francisco to hold his granddaughter as soon as he could, and then he headed to our apartment for a few days to cook meals for us to have when we got home from the hospital. My mom would come two weeks later from New Jersey. There was nothing left to do but lie back in the hospital bed and rest. After so much activity, so much preparation, so much multitasking for work while waiting for my daughter to appear, here she was. The only job I had now was her.

Plenary

I'll never forget my first conference after having our daughter. I'd been invited to give a plenary talk. It was the primary annual meeting of my professional organization. My husband and daughter came along, supported by a dependent-care grant from my university. She was just shy of fourteen months; I was still nursing her and wasn't ready to be away from her with a pump as a substitute. It was in Seattle, so Steve wanted to see some old friends too. Yvonne and her husband were still there, and she hadn't seen Garland-Rose since the short weekend trip she'd made to keep me company when Steve went out of town for a friend's wedding. She basically knitted and read while I nursed Garland-Rose and slept when she slept. That was my life.

I was also giving a talk on the hydrohalite work. Before giving birth I'd been able to wrap up and analyze the simulations of the possible climates of planets receiving different amounts of light from M-dwarf stars. I'd run simulations where I hadn't changed anything about the type of ice that formed on the planets. When temperatures reached below the freezing point of liquid water, what formed was regular, salt-free water ice, with the albedo associated with it that the field had come to expect since our early work when I was a graduate student. I called these the "control cases." Then I'd changed the albedos of the ice to match how reflective ice would be if it got cold enough for salts to crystallize in it from the sea water, forming hydrohalite. Hydrohalite was much more reflective than regular water ice in the infrared—the part of the spectrum where M dwarfs emitted strongly. So I made the albedos much higher to match this surface type in the model setup, when temperatures reached below a certain level—the level where hydrohalite formed in laboratory experiments my colleague Regina and others had done. This was about 20 to 40 degrees lower than the freezing point of liquid water. I ran the simulations again for these "hydrohalite cases." Each simulation took about one to two weeks to finish. Then I had to check that the simulations had reached equilibrium, meaning the climates on these hypothetical planets were relatively stable, with no major temperature differences above a degree or so from year to year. Once I'd verified this for all of my simulations—and there were dozens—I dove into looking at the results. I'd compared the climates of the control and hydrohalite cases. There was a difference. And a pretty big one, depending on how far away from their stars I put the planets. We'd discovered that even planets orbiting close enough to be in the habitable zones of M-dwarf stars could be colder than originally thought, because temperatures somewhere on their surfaces got cold enough for salt to crystallize and form this bright, highly reflective hydrohalite crust at the top of the ice. It was an exciting new result, and it meant that climate models needed to start incorporating parameterizations for the formation of this surface type if we wanted to get the most accurate assessments of what the climates were really like on exoplanets, and how habitable they actually were.

I didn't go to many other talks except my own. This was unusual. The main reason to go to a conference is to catch up with colleagues and former classmates you haven't seen since grad school, and make connections that might lead to new research opportunities, papers, jobs. This conference was different. I was there with a baby. Steve was there, but he wasn't a robot. He had friends to visit too. Our nephew had just started college in Seattle, and we wanted to check in on him and give him a good meal. And I didn't know how to turn a blind eye to my daughter's needs. *And* what she needed most, at just shy of fourteen months old, was me. Justifiable or not, my priorities were clear. Get in. Give the talks. And get out and enjoy being with my family and the old rainy city we had grown to love and found that we missed. From our premiere high-floor suite with floor-to-ceiling windows, we could see for miles, and we surveyed the landscape of dark mountains and deep green forest from the downtown metropolis.

I worried about how much I'd changed since I'd become a mother. People asked me about myself—my job, my history, my aspirations. I answered with one side of my mouth, stared at them through one eye, gestured with one hand. The rest of me was on her, focused so deeply into her ocean. I swam in her eyes, her glance, her mouth chewing on a small, soft pea. She popped them like candy. My healthy girl.

My life had been taken over by this creature. This extraordinary, magical, mystical human who had grown and developed inside my body for nine months, then appeared outside of my body with the generous help of medical professionals, my husband, and a trusted friend. I'd spent so many years thinking about life—where it might begin on other planets, what it might look like, how it might evolve. Then right here on Earth, I'd had a front-row seat to that process. Through imaging technology I'd watched her evolve as she grew arms and legs and a face. She had emerged, fully formed, with sparkling blue-grey almond-shaped eyes, the crinkly face of an old woman, tiny toes and fingers, and a huge head in the ninety-seventh percentile. Over the months she had filled out, her face plump, her arms and legs becoming little Pillsbury Doughboy shapes, and her smile—*God that smile*—that greeted me every morning as I peered over into her bassinet to meet her waking eyes. Her smile moved the Earth beneath me and made the morning something I wanted to walk into.

I didn't realize how consuming it had become until we started taking her to daycare a few days a week for a few hours a day, just to get her used to other caregivers as we slowly ramped up to the fall, when I would return to work full-time, and to teaching. It was a sane, well-thought-out plan. It would give me some breathing room. A few hours, several days a week when I could do something for myself, whether that was work, or nap, or get a massage or a mani/pedi. I knew it made sense, and part of me welcomed it.

One of those early Fridays, I walked to campus with my lunch, and instead of scarfing it down at my desk before getting down to some serious work, I found an available park bench and ate outside in the afternoon sunshine. I couldn't remember the last time I had eaten outside, alone, with two hands. I was used to eating in front of the TV, food my husband had prepared for me, while nursing our daughter and grabbing forkfuls of food with the other hand, trying to watch a show like we used to, before her 6:30 bedtime. This meal on the bench started to feel like the beginning of a return to something akin to who I used to be, and I felt giddy.

On another day, sitting in Tender Greens waiting for my grilled salmon, I didn't feel giddy or excited about being out on my own at a restaurant. I felt only a rift. Who was I now, without my daughter physically attached to my breast or my arm as I carried her around the house from changing table to baby carrier to play mat to bouncy seat to swing to nursing pillow to bassinet, and repeat? How was I supposed to switch back and forth from the all-consuming role of mother to an individual working professional who does research and gives illuminating talks and responds to emails from colleagues in my department and from people around the world who are conflicted about their careers and want advice? I'd submitted the hydrohalite paper while Garland-Rose was nursing on my lap. I pushed her in her stroller all over our neighborhood and napped with her in my arms and propped up on pillows. How could I go back to being the person I was before she was born?

It dawned on me as I shoveled a forkful of seasonal vegetables into my mouth—I couldn't go back to the way it was. And I didn't want to, for that matter. There was no going back. There was only going forward.

I knew I should be separate, apart from her in spirit, my own person. But it was lip service only. In practice, it was impossible, like making the Moon stop revolving around the Earth, the Earth around the Sun. I wanted to make it okay that I'd changed this much, that I didn't like to talk about what I wanted for myself apart from her. Maybe it wasn't okay. Maybe it just was. And that was okay.

What I remember most about that trip was the plenary. Not because it was a plenary, but because of something I heard myself say during the talk. In saying it, I claimed all that had changed inside me since becoming a mother, and embraced this new component of my identity for the first time in a professional space.

A plenary is a talk that is given when nothing else is scheduled at a conference, so everyone can attend. And almost everyone did. They opened up two adjoining ballrooms to accommodate the crowd. Yvonne came, as did my PhD advisor, Vikki, and colleagues from throughout my life. It wasn't because for the first time ever I hadn't rehearsed my talk in advance, which was terrifying right at the end. Up until an hour before, I hadn't had time to think about it. I'd taken Garland-Rose in a front-facing baby carrier strapped to my body to the speaker audiovisual room the night before to upload the final version of the slides for my talk. I'd been proud walking through the hallways and up the escalators, meeting everyone's warm and wonder-filled gaze with a smile. This was also unusual to see at a conference—an attendee with a baby in tow.

But the most unusual was yet to come. After I kissed Steven and Garland-Rose goodbye and headed to the conference center; after I found a nearby conference room and did my requisite vocal warm-up (I might not have had time to rehearse beforehand, but I was damn sure going to get my voice ready for the thing); after the introduction, by the very woman who, long ago, as program officer for the NSF postdoc fellowship I'd been awarded, told me that they wanted to find the solution that worked for my whole life; after I started talking and fell back into my old self, with a decade of knowledge under my belt about these planets around other stars, and all of the ways, large and small, that we were learning more and more about them, leaving still that much more and more to learn, mid-sentence I heard a squeal. It was unmistakable. Garland-Rose was somewhere in the room. I knew that squeal anywhere. High, pure, happy, like she'd just discovered her favorite, long-lost toy. She had. She'd seen me up on the stage.

I exclaimed automatically, into my lavalier microphone, "That's my daughter!" I scanned the back of the ballroom until I found her, in Steven's arms, dressed in her black-and-white striped dress that I'd laid out for her that morning, with the pink sleeves and the planet Saturn embroidered on the front. She was also wearing a badge she'd been given at the registration desk that read FUTURE ASTRONOMER. I waved at her and said, "Hi!," beaming out into the crowd. She jumped up and down in Steven's arms. I said one more thing to her before Steve took her out to run around in an empty conference room down the hall. "Mommy will be done soon."

Something else unusual happened while all this was going on. The audience applauded. Some people cheered at the "Mommy will be done soon" comment. There was no going back now. I had given the most important talk of my life. The professional and the personal had overlapped, blurred edges, intermingled like wine and cheese. For those few minutes, I turned the world of professional astronomy on its head, without the need to discover life on another planet. It appeared that many in the crowd liked the view.

NEOWISE

One of my new graduate students asked me where in town was a good place to view Comet NEOWISE. I hadn't heard anything about it. I'd been avoiding the news. We were in month five of the pandemic. Turning on the news was risky if you wanted to stay in a good mood, if you were lucky enough to be in one in the first place.

When I read up on it, I got excited. It had an extremely hyperbolic orbit. And it had slingshotted itself to within twenty-seven million miles of the Earth, closer than the average distance Mercury is from the Sun. It could appear bright in the sky. This was the first time since my daughter had started talking, understanding things like what planets were, that something like this had happened. At two-and-a-half years old, she now knew the names of all the planets in the solar system. I'd brought home the planet plushies I used for Rising Stargirls Workshops. They were hers now. Her favorite one was Neptune. She even had a gray comet plushie, with a poofy tuft of white fluff forming its tail. Real comets look this way when they get close enough to the Sun for its flames to heat up the ice and turn it into a long, gaseous trail. NEOWISE would look like this.

The best time to see NEOWISE was an hour to an hour and a half after sunset. This was summer, so that meant about 9 or 9:30 p.m. My daughter's bedtime was between 8 and 8:30. I told Steve about my dream of showing our daughter her first comet in the sky—a monumental milestone in her budding astronomy education. He reminded me of her bedtime. My mind swirled with *Yeah, but this is a comet! It won't be visible again for seven thousand years! Surely this is worth her losing a little sleep!* I had forgotten my responsible mother role. I was an Astronomom. The sleep-filled nights we were finally enjoying, the committed bedtime routine—I was willing to give up all of it. This was *space*.

Moons and planets can lock into specific orbits so that one side stays perpetually in the light and one side perpetually in the dark. I've told you about this type of rotational state before. Our Moon does this. So does Iapetus. But moons and planets aren't the only bodies with a permanent night side. I began to understand the feeling myself when I had a baby.

I saw every night time on a clock's face after my daughter was born: 11:34 p.m. . . . 12:11 a.m. . . . 1:23 . . . 1:48 . . . 2:15 . . . 3:26 . . . 4:17 . . . 4:58 . . . 5:30 . . . 6:38 . . . I would get up to nurse her back to sleep. I would get up to rock her. My husband rocked her more than I did, trying to preserve some fraction of the sleep he knew was more precious to me than emeralds, and far more essential.

After Garland-Rose had been tucked in, read and sung to, kissed goodnight, and left to fidget and fondle her stuffed animal bed friends until she surrendered to the pull of gravity on her eyelids, I occupied the quiet space between awake and asleep. Steven took the monitor downstairs with him to the online poker game he'd played since the beginning of the quarantine. In this little pocket of time, no one needed me—for sustenance, comfort, or companionship. I was alone.

It was a liminal space. I'd signed off work and not yet relaxed into the book on my bedside table. This was the time when any thought could grab hold of me and tug, pulling me down and wrapping me around a tree. One glance at a photo online of the back of a Black girl's head, full of barrettes, and I was thinking of my daughter and her hair clips that I pinned to her hair most days after braiding it. Her childhood stretched out before me, an unknown road whose twists and turns I couldn't anticipate. I squirmed in bed. I thought of what I'd done or not done right that day. I thought of the future and wondered if it would be good. I thought of playgrounds and tetherball and friends and bullies, and I wanted to shield her from everything that could hurt her. I still remembered everything that hurt me. The white boy who told me that he liked my best friend because she was white, and that he didn't like me because I was Black. How I told the teacher on him, and she made him come inside from recess, but it didn't make me feel any better. My daughter would pass for white in some circles, and I started to think that maybe that was good. Then I remembered how happy it made me when my friend Susan texted me after seeing a recent picture of Garland-Rose and told me how much of me she saw in her.

I thought of all the years of youth, and how we were in it with her for the long haul. I wanted to skip ahead, have her jump over those years of pining and pain like a rushing, ice cold river, so she could look back behind her to see it over her shoulder as she placed a sure foot on the solid ground, heading on to better things.

I tried to remember what Susan had said—that my job wasn't to keep her from experiencing painful feelings, but to show her how to handle them. Mostly to just be there, through it all, without trying to fix it or take it away. That sounded as easy as standing inert next to a shotgun as a tiger stalked and devoured a lamb. Simple and easy aren't always the same thing. *Fuck the circle of life. I must save that lamb.* I am her sentinel, and no tigers are getting through on my watch.

In couples therapy, the difference between Steven and me is obvious. Steve knows we're doing our best, making mistakes along the way. I ask questions, one after the other. Eventually I'll have asked them all and written all the answers down so I can unlock the secret to ensuring the perfect life for my daughter. No pain. No loss. No neuroses. All confidence and an open field to wander in, free. I keep asking. And in this space alone—no her, no him—I can keep asking forever, and all night long. And I need to sleep. This is why it's hard to make time for myself. That it should go to thoughts of someone else feels right and just. There is so much to do, read, read to do, read not to do. I haven't been paying hard enough attention. Something is slipping through the cracks. I know it.

I stretch my thoughts as far as they can go. The taut rope finally relaxes. It is a hard lesson, but it helps when I remember that this is just one day. I don't want to only deal with today. I have someone's lifetime to worry about. As a professor, I'm expected to know all the answers. So of course I want to know them all at home too. But that approach won't get me any sleep tonight. And I love my sleep.

I love my work too. Most of the time. The possibility of quitting my job never entered my mind, in all my decades of working. Until I had my daughter, at age forty-two. That was the first time that I realized just what work demanded of me, just what kinds of sacrifices I would be making to have a professional career. I cling to my values. I will do her hair in the morning, even if it means starting my workday later than most faculty. I end my workday early, so I can give myself time to unwind, so that I can give myself the best chance of being a happy, serene person when my daughter gets home. This will help me, and by extension, my husband and daughter. I will take care of the self, so that I don't lose myself. So that I don't bank my sanity and peace of mind on a potty time or a bedtime or a play date or any particular outcome.

I will spend that hour with her before dinner—not on any phone, not dashing off one more email. I will read her bedtime stories, sing her songs, and tuck her in at night. I do all of that imperfectly. And none of that ever feels like enough. Being a good mom means more to me than being a good scientist. I have communicated this to Steven, and he understands now. He knows that he lives in a catch-22, where his increased involvement since weaning her from my breasts is simultaneously appreciated and resented. And he knows I'm working on the second part.

Though I was antsy about the possibility of the NEOWISE viewing opportunity disappearing into thin air, the next thing presented itself—dinner. Then bath. Then lotion and pajama wrangling. We told her to sit on the potty. She did, without complaint ("no guff," as we like to say). Garland-Rose hadn't pooped all day. We were in the throes of potty training, and while peeing in her potty was going well, the idea of defecating anywhere other than in a nice, soft, cushiony place right up against her bottom, where she didn't have to watch feces actually exit her body, was still off-putting. To the point where she preferred keeping said feces inside her body instead of depositing it into a toilet. While we were not wild about her fad of pooping in her underwear, her not pooping at all was far less desirable.

Less than two minutes later, I watched her look down into her potty and say, "Yay!" I sidled over to peer into the bowl. Inside was a single, long, perfect turd, in shades of cocoa brown and deep leafy army green. It was one of the most beautiful things I'd ever seen. It was so perfect, so cleanly deposited from my daughter's anus that when I wiped her, there was hardly any poop on the tissue. It was sublime. My husband said it best. "It was our own little NEOWISE!" Indeed.

NEOWISE

I let my daughter go to bed. There would be other comets to see in her lifetime. I trudged up to the lookout by myself, with my huge pair of binoculars to take a look. I saw a white, fuzzy blob with a wispy fan of a tail. She never would have been able to see it with her naked eyes. Maybe if I'd had a tripod for these binoculars, I could have convinced her to look through. It had worked out the way it was supposed to.

RISING

I Never Thought I Was Only Located Here

I always knew about something bigger,
beyond where my feet are,
my hands, toes, brain.
It started with the craning of a head upward.
Had my neck muscles not fought against the position,
there I would have stayed, looking skyward.
I wasn't only looking. I was waiting.
I waited for the stars,
the breath in the black,
to exhale its secrets.
I waited for the sounds and watched for the sights of my
    kindred.
Up there,
beyond the gas and the dust,
the atoms and molecules,
the stars bursting into brilliant flame,
forging out through great swaths of space toward eagerly hungry
    daughter planets,
there is a home.
It is my home, of a different nature.
Beyond the sheets and toys,
the alcohol mindfulness and the water consumption and the
    television,

the meetings and the marriage and the bedtime stories and
    songs,
something else exists, and I can feel it.
My eyes strain to see.
I know I need the most powerful telescope beyond the tallest
    mountain peak
and deeper into space than ever before,
with more mirrors and elaborate intricate petals unfurling.
Still, it would not be enough.
That is all science. This is intuition.
It is the underbelly of why I look,
why I work and write papers and speak.
It is a sureness of footing,
a deep-seated feeling
that we are unequivocally, assuredly not alone.

Aliens

I am looking at a picture of a woman crouching on a rock in the shallow waters of a lake in Norway. I think it is a woman, though she is in shadow. The picture was taken in low light, maybe at dusk, from inside the canopy of a dense forest. The picture is in my desk calendar, and it makes me think of something bigger. I can see that the lake snakes around a small island of trees, and beyond it lies the curve of more land covered in dark green trees. There are a few logs at the lake edge, covering a bed of fine silt and stones. A stocky rock close by juts out into the water. The woman is looking to her right across the close surface of the lake. The water isn't completely still. There are patterns in it, ripples, tiny imprints of a rhythmic wave front set in motion by something. Maybe her feet in the water, maybe the swift dive of a passing eagle grabbing trout from the upper inches of the water. Diffuse light from the Sun has caused the glare in the photograph, extending a white carpet of brash light across the lake.

In a turning moment my thoughts go wide. All of these planets that we know are out there. Billions. How many of those might have—do have—a lake like this, surrounded by a forest, full of thousands of species? And perhaps there is a woman standing or crouching in the water on at least one of those planets? Does she herself wonder how unique or common she is in the universe? Does she struggle? Does she experience the joys, the pain, the sorrows of life just as we do?

It all gets too big in my mind. The implications! I know this when it becomes hard to find Garland-Rose in it all. The more I think of all the planets in the galaxy, let alone in the universe, where this scenario might be playing out at this very moment—a woman in a lake in a forest full of life, the sheer grandeur and beauty and abundant richness of it all—the smaller my daughter gets in the picture. Where is she? She becomes a speck on a CCD frame. Too small, even. She is one single drop in an ocean teeming with life.

I have to come closer, back in, to this planet. I do. I stay with the photo of this one woman, in this lake. My thoughts move again, this time to a different unit of measurement. I think of long ago, of how this lake has sat still and relatively unchanged as life has filled it, died, and new life has filled it again, and again, over and over, for millennia, and before that. It's like the book I read to her at night, about the stone that sat still and was everything for everyone, and was also a memory.

Again I lose her in the unit of time. My daughter wasn't there with the lake long ago. She wasn't a thought, or even a thought of a thought of a path that might lead to a thought, on an ancestral family tree, somewhere in a far, dark continent. She was yet to be in anticipation of. Anticipation itself was yet to be. Could life have existed before she breathed it? I know that it did, but, really, could it have?

The only place I keep her alive and strong and present is now, in the picture. But not even the picture itself, which was taken in 2013—four years before her birth. It is in the seeing of the picture, me seeing it now, taking in its light from page to eyes. In this moment—of my understanding, contemplating, wondering what can and maybe does exist and has existed, without diving deep and irrevocably—she is, and I am. And I can breathe again.

Why does it matter whether life exists on another planet? It matters in the same way that a mountain matters and screams to be climbed, to those whose ears are sensitive enough to hear. We have to know. I have to know if we are alone. I can't stand not knowing, if there is something I can do about it. Sometimes, most times, there isn't very much I can do about the world, the state of affairs, even the day-to-day. I can send money to organizations that do good things. I can join the picket line. I can join the strike. I can write an op-ed, take a meeting, use my power and influence to get that Black student with untapped potential into my graduate program, knowing in my bones that she has what it takes and hasn't been given the chance to show it in a nurturing environment that allows her to bloom like the sunflowers in my backyard. But beyond that, I have to accept the legion of events and unknowable actions that are beyond my power to control.

Not so with the search for life. While I'm still here, still breathing, and still interested, there are things I can do to answer this question humans have been asking since they began scrawling pictures on the walls of caves: Are we alone? Is it just us, or are there others out there? It is a question humans asked about the other sides of the world. Answering the question, motivating the voyages of Columbus, Magellan, and countless followers, led to devastation and genocide. The answer wasn't enough. What followed, perhaps what inevitably follows, is conquest, the desire to control what cannot and should not be controlled.

I say I have some control here in finding out whether life exists elsewhere. I use the data I get from telescopes, about where a planet is around a star, as a springboard. I assume something about the planet. Maybe that its atmosphere is like the Earth's, to start. Then my computer simulation begins. Starting with its distance from its star, and having the atmosphere that I've assigned to it in the computer code, I can tell if it could be warm enough for surface liquid water—that all-important elixir and ingredient for all life on the Earth. We want water on the surface, rather than below an ice sheet or a kilometer of dense granite. The closer life is to the surface, the closer life's exhalations are to our telescope's instruments, staring into the vast blackness of space, straining to get one pixel of gold, to produce one signal that will change the fate of our time.

If a planet couldn't have liquid water on its surface with an Earth-like atmosphere, then I start turning knobs. How much thicker does its atmosphere need to be? The thicker it is, the more heat will be generated. We have to be careful though. Too thick a recipe, and we get Venus. Too thin, and we get Mars—with an atmosphere so tenuous that water couldn't flow on its surface today. If we landed on Mars, in our thick space suits, and opened a pressurized canister full of fresh water and poured it out onto the rusty ground, it would likely evaporate within seconds.

This is what I can do. I can tell you whether the planet you found has a snowball's chance in hell of supporting life as we know it. I can tell you if your planet could support that life easily, without batting an eye, no matter what kind of atmosphere it has—thin, thick, filled with nitrogen or carbon dioxide. No matter what kind of surfaces are on it—ocean, desert, water ice, or some other kind of ice. No matter what the shape of its orbit or how tilted its axis. No matter how active the star, what kind of star it is, and whether the planet's plates shift like jigsaw puzzle pieces. If the planet couldn't complete its mission of succeeding as an easy home for life, I can tell you what it would need, what best recipe combination would get it there. Or I can tell you to move on, this planet is a lost cause, best to point your life-seeking telescopes elsewhere. And if you're looking for life as we do *not* know it, you'll need to ask someone else about that.

To know. To find out that we aren't alone. It would change everything. All our little, petty worries and concerns. I don't believe I will truly accept or understand how big the universe is until I know that other lifeforms share it. With that knowledge, the sky is no longer a snow globe, but a window. The universe, the galaxy I peer out into at night, peers back at me. Every sci-fi series and movie I've ever seen—from *Star Trek* to *Contact*—becomes real, no longer fantasy. The universe will have opened its secrets to me. Not only are we no longer alone, but also, no longer am I—one tiny individual who *must* look up at the sea of stars and wonder—alone. Something is looking back at me.

That's the *me* part of why it matters. There's also, now, the *her* part of why it matters. Everything I tell my daughter—how many stars are out there, how many planets orbit those stars—means more, matters more, because those stellar systems are not barren. They become homes, like hers. They can speak, they have histories beyond what we can measure through a telescope. They live and breathe in whatever way they have learned. They tell bedtime stories, maybe. They become *they*. Not because they choose to be, but because they *are*. They exist. They have a story of their becoming, and one day, my daughter may hear it. There is a future for her that is different from my past, where she exists in a universe with multiple origins of life. The outcome of that difference is unknown and cannot be controlled. What is known is that her horizon will expand. I could never regret her knowing that.

☾

Oumuamua. It sounds like my name. It means "a messenger from afar arriving first" in Hawaiian. I didn't know about it when it swooped by our Sun in 2017, on its long, hyperbolic trajectory. The first object to visit us from another stellar system, with its own parent star that people are still trying to identify. A lost child in the big city. To whom does it belong? The parent star must surely miss it.

And so strange. The object was oblong, a rock spear in space, like nothing—no asteroid or comet—we'd ever seen. We knew that not everything in space was round. Some things are big chunks of rock and ice, shaped like crumpled-up pieces of paper. But a *spear*. A spear strains and pulls at brain cells to figure out how a shape like that would occur naturally. On this planet, the tips of spears were made by Indigenous people from bone, obsidian, flint. To make a spear required more than raw materials: it required effort and consciousness.

Everyone has a theory. A hydrogen iceberg. A nitrogen iceberg. It shifted its trajectory. Naturally occurring astronomical objects don't do that. The word *phenomenon* began to take on new meaning. And the loudest theory shouted from the rooftops was that this object was not a natural phenomenon, but one of alien origin. Alien technology.

As much as I believe it to be unlikely that we are alone in the universe, as much as I want to believe in life beyond the solar system, and in UFOs, Roswell, conspiracies, and cover-ups, I am a scientist. And since I am a scientist, the line between my beliefs and actual fact must be clear—as clear as the sky after a torrential storm over the dark sea has passed. The line is clear and firmly planted in soil. I will not move it one inch toward my subjective beliefs, hopes, desires, and dreams, for anything.

The *I want to believe* poster I had on my wall as a teenage fan of *The X-Files* remains exactly that—a poster I once had on my wall as a kid, and a feeling I admit to you, with no basis in empirical data or other physical evidence. I would love for Oumuamua to be the first visitor from an alien species. I would love it even more if Oumuamua wasn't the first, or the twentieth, of such visitors, but one in a long line of visitors to our solar system over millennia, watching and waiting for us to be ready for this knowledge. Then every movie, every television show, every science fiction story I've lost myself in over the years will have been prophecy. But while I dream, science is being done.

I follow what the evidence tells me. I listen when scientific explanations are found for previously unexplained phenomena. I listen when a paper comes out that finds that sunlight interacting with nitrogen ice can explain what was observed. Nitrogen ice reflects two-thirds of the sunlight that hits it, so it can stick around for hundreds of millions of years in interstellar space. Nitrogen ice also outgasses, which could explain the accelerations in this object's trajectory, and since scientists weren't looking for nitrogen ice, they didn't detect it. If new explanations are revealed to refute current ones like these, I'll listen to them. And if the best, most convincing, data-based explanation is an alien origin, I will listen then too.

I truly believe that when the day comes for us to discover that we are not alone in the universe, we will know it. We won't have to guess, or theorize, or even debate it among our scientific communities. It will be made as clear to us as the sky after a torrential storm over the dark sea. The line dividing the time before humanity possessed this knowledge, and after, will be clear and firmly planted in the soil, like a flag on the Moon. We will know where we've been, and where we are in it all. A moment like that would not present itself as an afterthought with an extant, untraceable source. No riddle will need to be solved. We won't have to believe. We will unequivocally know.

Congratulations, Professor Shields

As a new professor, tenure is the trophy, the Holy Grail, the magic bean that, when you've been given it, makes you free. Not free like Underground-Railroad free, but a free that you feel in your bones. The tight constraints of academia, the words kept close to the chest in faculty meetings to avoid repercussions years down the line when senior faculty get to vote on your case, can finally be loosened without danger of revenge or permanent damage to your career. In layperson's terms, it means you can never be fired. Permanent job security, barring something really bad (like sexual harassment—but sadly, even that might not get you fired; murdering someone though, probably would). In today's post–COVID-19 economic climate, it's job gold. It's a winning lottery ticket.

Given the grandeur and high-impact value of tenure in academia, attaining it is naturally a process that requires a great deal of time and effort, including publishing as many papers as you can write or have your students and postdocs in your group write. (They carry equal weight.) The necessary balance must be struck between doing enough service to be seen as a contributing member of the department, but not so much that you neglect to turn out the papers needed to be labeled a true scholar in your field. Good, consistent teaching is required, but without spending too much time working on being a master teacher that you neglect your scholarship. At a Research-1 institution, great teaching and a lot of service without a high level of scholarship (aka "research productivity") will not get you tenure. Excellent scholarship without good teaching and service probably won't either, but it's certainly going to get you closer, and is often considered the more forgivable of the two scenarios.

I had the option to stop my tenure clock to put a pause on the review process. It would have meant that I'd have eight years to get tenure instead of the standard seven. After all, I'd had a baby—a perfectly acceptable reason to do it. My coach and I discussed it. I'd won two major grants during my first year as a professor on campus. I'd published a paper and had an extraordinarily productive year, even for someone who hadn't also grown and given birth to another human being in that same time. So, I didn't stop the clock. Instead, I returned to work full-time the following fall, returned to teaching, hired a postdoc and a grad student, and got them both working on papers on the effect of land fraction and host star spectrum on planetary climate, and the effect of eccentricity and host star spectrum on planetary climate, respectively. And I wrote two more papers myself, one calculating the effect of host star spectrum on the energy budget of an orbiting planet—how a planet distributes the energy it receives from its star throughout its atmosphere, surface, and back out into space. The other paper was a sole-authored review paper on the field of exoplanet climatology. It was a way to look back at this new field I had been working in for the past ten years, starting with how climate models were used first for the Earth, and then other planets within our solar system, and then planets orbiting other stars. It had been gradual, and there were still only a few dozen of us doing it. But reviews like this would spread the word so that there would be more.

Regular merit reviews were entirely internal and based on the judgments of faculty, chairs, and deans at the institution, based on the CV, statements, teaching evaluations, publications, and grants of the professor. For tenure, it was a whole different ballgame. All of those documents were still required, and they had to frame the professor's work in the larger context of the field. But to build a successful case, the professor needed to be regarded as a pioneer of something by those in the field and the larger scientific community. So the professor's case needed to be backed by five to ten letters from colleagues outside the professor's institution who attested to the innovation and pioneering brilliance of said professor, and swore on a stack of Bibles that they felt said professor's work over the past six years had more than earned them tenure at their institution. The entire case, including statements, teaching evaluations, number of publications and grants, and excerpts from the external letters, was presented in a department faculty meeting where the professor was not present. Then an online vote survey was circulated by the chair's office. If the vote was positive, the case was advanced up the chain, to the dean's office and on to the Committee on Academic Personnel, and then sent to the provost and chancellor for final approval.

I didn't used to know much about this process. I wasn't planning to learn any of this detail until at least my fifth year, after my fourth-year review, when I expected to receive some useful feedback about how I was doing, whether I was on track for tenure in year seven, if I needed to say no or yes to more service, if the number of publications I had at this point was good or if I needed to step it up.

My department had a history of turning the fourth-year, typically midtenure review into a tenure review—essentially fast-tracking particularly deserving candidates into getting tenure in half the typical time. This was largely unprecedented in other departments on campus, let alone at other institutions. I had voted on one or two of these "early" tenure cases since I'd arrived at UCI. I hadn't given it much thought beyond checking a box in an online survey for someone who seemed to deserve it and, as indicated in faculty meetings where their cases were discussed, seemed well-regarded by their peers in externally solicited letters.

When it came time for my fourth-year review and I was encouraged to follow a similar strategy, I was shocked. Wasn't it too soon? Even though I'd seen it done with others, this was different. It was *me*.

☾

I tried to operate as if my professional world hadn't just been turned upside down and given a good shake. I tried to inject some balance into things. *Right, I have a personal life too. Let's get back to that. What is my husband doing right now? What would he like to be doing instead? . . .*

For most people, the idea of sex in the afternoon on a weekday would sound like a grand idea. What a naughty, delectable thing to do, like playing hooky. Steven told me that the blue-and-green tank dress I'd worn the previous night, that hugs my body and won't let go, had driven him mad. As we'd predicted, what everyone said was true: finding—perhaps making—the time for sex had indeed proven to be more difficult with a child. But the attraction was still there, and as my body had begun to feel more and more my own again over time, I appreciated it when Steven let me know that I was attractive, that I still "had it." Wednesday was our real date day, when I would take a full four hours off to go sit in the darkness of a high-end movie theater with him alone at 11:30, no one else in the world around. The idea of relaxing and watching a movie and having food brought straight to our recliners made racing home to pack sex in on top of it all less appealing.

So here we were on a Monday, in our bedroom at 4:15 p.m. One of us had to pick up Garland-Rose from daycare at 5 p.m. Steven had been out all day playing golf. It was a day off for him. As our bodies moved, as my lips met his and entwined rough against smooth, briskness of unshaven face, firm hands holding me tightly then releasing to let me move and shift and mount and then submerge, I thought of so many things. I realized I hadn't brushed my teeth that day: working from home, I'd showered first, then eaten breakfast and never made it back to the sink. I thought of the email I'd sent to the journal asking, in more polite terms, what the hell the holdup was with my review paper, the latest version of which I'd submitted over two months ago. I thought of the rapid-fire email apology from the editor, and of the next four emails that followed, informing me of the paper's acceptance, and proofs to come, and next stages. I thought of the email telling me that my mid-tenure review was that fall, and of the follow-up email from the chair letting me know that I could consider going up for tenure instead, and we could discuss it if I wanted to. I was still thinking, *Holy shit, holy shit, oh my God, it's too soon.* But I didn't want to undersell myself. A man, a white man for sure, would be like, *Fuck yeah, put me up for tenure early. I'll take those odds, I'm a BOSS.* So what the hell, right? Or wrong?

I was thinking I would have to discuss this with my coach before I met with the chair the next day. I had a fucking kid, and maybe there was no reason to rush tenure. But the chair said it wouldn't be considered an acceleration. *Oh my God.*

All of this was going through my head while Steven and I were having sex. He asked me if I liked it, how this felt, our bodies against each other. *Uh oh. He knows I'm divided. He KNOWS.* I deflected, saying, "Are you kidding? I wouldn't be here if I didn't like it." I was failing. Thich Nhat Hanh's book *How to Relax* sat on my bedside table. *Breathe, and be in the present moment.* This was the first rule.

CONGRATULATIONS, PROFESSOR SHIELDS

I hadn't gone looking to get tenure so soon—but I was told that it would be a "slam dunk." So I started to dream and bolster myself, talking to my coach and my mentors, on campus and off campus, and I thought, *Hell yeah, bring it on, I can do this, I deserve this, I've won this grant and that grant, and I have an organization that serves middle school girls of color all over the world, and I'm publishing like gangbusters now and my team is on fire, we're three strong and we've got five papers submitted in one year, count 'em, and beat that, motherfuckers, hell yeah it's a slam dunk, you said it, now you made* me *believe it!*

The stakes for tenure may be lost on people who have always had money, or have always had a job or a paycheck, but they weren't lost on me. With tenure, I would always have a job. With a paycheck. A six-figure paycheck. I would be paid continuously, for the rest of my professionally active years, until I retired. I've always been educated, gone to the best schools—yes. That in itself is some degree of privilege, even though I'm a Black woman and a member of two marginalized groups. There's nothing marginalized about my résumé. But I have not always had money. Working as a temp at that music company for a year and a half, with a bachelor of science degree from MIT, I was angry at times. I knew what it was like not to be able to pay the bills. Never again.

I wasn't prepared for the backlash of my tenure bid. It was not at all a slam dunk. Half of the external letters praised my work as "pioneering" and "forward thinking." The other half carried a lukewarm tone, punctuated by "Isn't this early?" and "It isn't done like this at *our* university." My personal favorite was a letter that said that my most significant work was my PhD dissertation, a backhanded way of saying, "She hasn't done enough yet." The same letter said that I had the potential to be the next Neil deGrasse Tyson. This kind of comment sounds good on the outside, but the letter writer knew what she was doing. Lauding me for my outreach and science communication was another way of telling the tenure review committee that my focus, my talent, my contributions were somewhere other than in the area that they should actually be to merit tenure: my research scholarship. It's why I'd stepped back from the TED community. As much as I loved the limelight and the work in communicating science in engaging ways to the public, I knew that wasn't going to get me tenure. Papers would. Grants would. With decent teaching and just enough service to be considered useful and collegial to the department.

I wanted to quit. *To hell with this*, I thought. *I wasn't even looking to be a professor in the first place. I certainly hadn't planned to go up for tenure early. I don't need this. There are other things I could do with my life that wouldn't make me feel like shit. Maybe I should just go back to Hollywood.*

But then I heard Natalie's voice in my ear. I saw her words on the page. *Continue under all circumstances. Make positive effort for the good.* Fucking hell.

I did the one thing I couldn't stand to do: I sat still. I let it all swirl around me. As the waves of anger, resentment, and deep hurt washed over me, I didn't turn away.

It all ended. The second faculty meeting came and went. The meetings with the chair of the tenure committee, the chair of the department, the dean of the school, the vice president of equity and inclusion both in my department and the larger university, approached and passed. The letters from the chair and dean, explaining that the external reviewers were asked to send letters so early in the process that they didn't know about three additional papers I'd submitted in the months since, were read. The vote happened; the reading in front of the chair of the negative comments that senior faculty, whose names I will never know, chose to add on top of their negative votes, finished.

CONGRATULATIONS, PROFESSOR SHIELDS

When the positive tenure decision was finally handed down, nine months after the initial documents were submitted and three months after the world had been shut down by a global pandemic, the letters from the chair of the Committee on Academic Performance, the vice provost, and the chancellor all praised me for my commitment to serving girls from underrepresented groups alongside my excellent scholarship.

I remember seeing the email in my inbox. And then others, with a "Congratulations" in the subject line—from the dean, from my department chair, from other faculty. I was sitting at my desk in my office at home. I stared at the emails, and I exhaled. Long and slow. For a moment, time—everything—was suspended. It was as if I could see the individual molecules in the air.

Then it all came out. I lifted my hands up to the sky, fell back against my chair, and screamed at the top of my lungs. Relief. And more. Later that day I painted all that I felt. On the thick watercolor paper there was rose and gold and purple and glitter and all of it swirling and fanning out, jumping out of the page. It was an explosion. It was the Big Bang.

Not long after that day, an email went out to Black faculty celebrating my tenure award. There are few of us here, so we have to celebrate these advances when they happen. One of the other Black faculty, a professor of dance I met when I first arrived on campus, said I had done it in "record time." The impostor voice came back as I read. *Maybe I didn't deserve the title, maybe I took a shortcut, cut corners, was judged more favorably in this time of Black Lives Matter.* But then I remembered the dates on the letters. I didn't get word until June 2. The letters from my institution recommending tenure were dated in April, before George Floyd was murdered on videotape. Before the world had finally had enough and decided that more should be done. In April, Black Lives Matter hadn't become fashionable yet.

Maybe I could choose to believe that the reason I had gotten tenure in "record time" was that I was outstanding. Like others before in my department, I had demonstrated exemplary research, teaching, and service in the four years since I'd started as a faculty member. Those words were reflected in the other half of the letters. I could choose to focus on those. I knew that I'd been doing the work of an associate professor for a while. This job could be, if I let it, showing up and doing the work, and trying not to judge the quality of it, like my friend Maggie told me a long time ago. Other people might and would continue to judge it, in many cases because it was part of their job. But it didn't have to be mine. No, it wasn't a slam dunk after all. It was hard fought. And hard won. I now knew how to be strategic in ways I'd never imagined before. I knew a lot more now. And isn't that always something to be grateful for?

First Day of Class

The first day of classes, September 2020. A new quarter. A new academic year. Under quarantine. The class I'm teaching—all on Zoom—is about planets, in the solar system, and outside of it. The course is largely in my area of expertise. This class has six graduate students. I have taught the class before. I have tenure now. And yet I haven't been this nervous since the first time I taught a lecture hall full of 130 undergraduates as a new professor at UCI. My coach doesn't understand why. She says I've handled teaching stuff like this before, no sweat. She's right. I'm not sure what the deal is either.

At first I thought it was because one of the graduate students taking the class took it the previous year, when it was offered as a special topics course (more free-form, almost like a seminar), and maybe I was bogged down in trying to make this course different for him, so he wouldn't be bored.

Then it dawned on me. It wasn't about the student. It was about the world outside of the classroom. It was about knowledge: how I love it, and almost as much, love to share it, and—here's the thing—I was not feeling very confident in how much of it I had to share amid everything that had been going on. COVID-19, the rise of the Black Lives Matter movement in the wake of the videotaped murder of George Floyd by white police officers, the same day as the videotaped call to the police on mild-mannered birdwatcher Christian Cooper by a white woman who knew the power she wielded with her skin and her gender combined. On the day I learned about both of these, a disturbing incident and a murder, I'd had no idea. An Asian female colleague in my department emailed me to see if I was all right, given what had just happened to George Floyd. I put his name in my internet search engine. And there it all was.

My plate was overflowing on its own. My daughter was still adjusting to preschool. Unlike the daycare "twos" room, where she got to do pretty much whatever she wanted whenever she wanted, now there was an actual schedule and instructions and things she had to do and places she had to be at certain times. Preschool was meant, after all, to prepare kids for real school. And all my daughter wanted and seemed to do was run out of the classroom door every chance she got, much to her teachers' Montessori-softened frustration. I was butting heads constantly with my husband about how to get her to do what we needed her to do at home on a daily basis, like go to the potty and put on her pajamas. I had my own health issues and weight gain as a result of being treated for hyperthyroidism. I had started writing a book—the one you're reading right now—most decidedly outside the academic job description for my faculty position. I was participating in an NCFDD workshop to discover and uncover what I wanted to do with my life post-tenure, which was buried very deep beneath the layers of perceived obligation and indebtedness that I somehow felt to the university for granting me the tenure and permanent job security I had earned. I hadn't had time to return any calls from friends, let alone stay up to date on the latest planet-focused papers that had come out that year.

With all that was going on out in the world and in my personal life, the topic of other planets and what was going on with them had gotten very small. There was enough to worry about right here on Earth. What if I didn't know anything about planets anymore? What if it had all left my head, the way basic information like what day it was did when I woke up in the morning? I'd been diving deep in my sleep, in my dreams. I would open my eyes and still be there. The world had to rush back in.

So I started with the Earth. I asked my students to tell me how we determine the Earth's temperature. The Earth receives light from the Sun, but that light hits the circular disk of the Earth that faces the Sun, like the shadow a ball makes on the pavement on a sunny day. A disk receives that sunlight across its area, which is the area of a circle. Do you remember from school what the area of a circle is? Pi times r squared. The r is the radius of the Earth, the radius of that circle that the Earth presents to the Sun. So it receives the Sun's rays across its projected area, and it uses some of that sunlight to heat itself. Eventually, the Earth is going to have to get rid of the sunlight it absorbs, sending it back to space, or else it will heat up too much. And the Earth doesn't need to get any hotter. There are still fires burning out of control twenty minutes away from my house, and people have lost their homes, their lives.

When the Earth emits sunlight back to space, it emits not just from its projected area—that circle—but from its entire surface, which is the area of a sphere—because the Earth is one big three-dimensional almost-ball in space, a sphere. And the area of a sphere is $4 \times$ pi r squared—basically four circles folded and wrapped tightly together to make that sphere.

The most fundamental principle of energy balance of a planet is this: energy in has to equal energy out. If those two sides aren't equal, the result will be a change in the surface temperature of that planet. I asked students to use that equality (energy in = energy out) to calculate the Earth's global mean surface temperature. Then I saw what they came up with. I hoped they'd started with the area of a circle . . .

But there would be a surprise ending. The answer they got using this simple method wouldn't be the actual global mean surface temperature of the Earth, which is 288 Kelvin. That is 15 degrees Celsius, and 59 degrees Fahrenheit. One of these numbers will speak to you directly, based on where you live in the world, and maybe what you do for a living. This is basically the average surface temperature in San Francisco, California. But the answer my students would get using their approximation would be 33 degrees colder—below the freezing point of liquid water.

They would wonder if they did the calculation wrong. The Earth is not a frozen wasteland! We have ice at the poles and around the world during the winter months in each hemisphere, but not constant, covering, eternal snow and ice and cold across the globe. Then why does this equation, once solved, yield a globally averaged surface temperature of 255 Kelvin? That's −18 degrees Celsius, and 0 degrees Fahrenheit. Where is that missing 33 degrees? What is not being accounted for?

It would take me months after being awarded tenure, sitting with the world outside and the world inside, to decide what I wanted to do with all the energy coming from both sources. There was an energy imbalance. There was excess heat from the world that I couldn't help but absorb. And I was adding to it the internal energy generated by how I processed and felt about what was happening outside of myself. I didn't know how to get rid of this energy while also keeping my hard-won job security and commitment to my own life and values.

Then a pathway emerged to help me begin to release that excess energy. I wrote an op-ed for *Inside Higher Ed* about how I had kept silent through the bulk of the tenure track when systemically racist practices were discussed in faculty meetings, and how it had gone when, tenured, I finally spoke up. It was my way of walking the picket line. And it was the start of a new way of being in this field. The original point of tenure had always been to allow professors the academic freedom to engage in daring new work without fear for the future of their employment. If there was a time to start saying what I wanted to say to save my own soul and hopefully help others in the process, it was now.

I would ask my students where they thought the missing 33 degrees came from. Some would say it came from clouds, which absorb sunlight. I would tell them that they'd almost gotten it. But they needed to think bigger.

Another student would offer the answer of radiogenic heat—the heat left over from the formation of the Earth itself. I love that both of these answers came from women—one white, one East Indian. The East Indian woman was a first-year student who would work with me officially as a member of my group after doing a summer research project with me the previous year as a rising senior undergraduate. She explored the way light interacts with carbon dioxide ice on cold planets—planets so distant from their stars and so cold that carbon dioxide gas literally falls from the atmosphere as rain and condenses into carbon dioxide ice on the surface. She'd gotten a taste of what it's like to use models to simulate the climates of exoplanets, and she wanted more.

I told the class that radiogenic heat is something else, and not very large—not large enough to cause a 33-degree increase in surface temperature.

It was time for the big reveal. What is large enough to cause a 33-degree increase in surface temperature? The greenhouse effect. I told them that the equation I had them set up and solve with only a few properties as input didn't account for a very important truth about our planet: the Earth has an atmosphere.

When the Earth radiates its heat from the Sun back upward toward the sky, it wants that radiation to make it to space. More is always on the way from the Sun, so it has to get rid of it. This is not the time to save.

The Earth needs to be a compulsive spender, or it will pay a price too high. It will burn.

Needs can't always be fulfilled. At least, not entirely. Some of the radiation the Earth emits makes it through the atmosphere and back into the cold blackness of space, in a sweet release. But some of it doesn't. Some of that radiation is too attractive for some gases to let pass through. They live for the kind of radiation the Earth emits. Infrared light, the kind we could never see with our eyes. The kind humans—you—give off all day long, just from your own body, living and walking and sleeping and eating and working and going to the bathroom. Carbon dioxide and water vapor in our atmosphere both love to absorb infrared light. Just like I, as a child, loved to absorb sweetened condensed milk by the spoonful. I didn't care that it was for a recipe for some other dessert. It *was* the dessert. I gobbled it up in the corner of the kitchen when no one was looking, and was happy.

So the carbon dioxide and the water vapor gobble up what the Earth sends up and away in an attempt to cool itself. Then these gases need to cool themselves. They respond in kind, and emit radiation up and away, as well as back down toward the surface. There is more than one direction in life. You know what I'm talking about. You can head upward, in the direction of your dreams, taking you higher and higher in a large rainbow-colored hot air balloon, your old, drab life getting smaller and smaller below. Or smaller and tinier in the rearview mirror as you speed up the coast in your cherry-red Mustang convertible. You finally made it out, and there's no turning back to him, her, them, that life before.

Or you can go down, and down, and farther down, the way it is during some years of our lives. The year in Madison, Wisconsin, when every choice I made was the wrong one and yet somehow I always thought it was the one for me, because it kept me from being exposed, and I thought that what was right was to hide. Walking to the grocery store knee-deep in snow, carrying four bags home, rather than asking for more than one ride from a classmate. Doing my problem sets alone rather than joining my classmates studying together in Astrohaus and risking their knowing I had no clue how to do those problems.

It's a good thing though, in its pure, basic form. Without that energy radiated downward, we would be frozen. We wouldn't be able to go to the beach or have picnics or take walks. We wouldn't *be*, at all. And, all other things being equal, the Earth would still manage to radiate away what it receives at the surface from those atmospheric gases, staying in energy balance, natural greenhouse effect included.

Unfortunately, all other things are not equal. As I finished explaining this to the class, as I saw my watch face spill a minute into the next hour and I knew I was overtime, there was something they had to know. This was no ringing endorsement for what's happening to our planet's climate. This is the fundamental principle of planetary climate and global energy balance. It is what everyone must know if they want to understand why and how planets support life, and why and how some of them don't and never will. We have too much carbon dioxide in our atmosphere now, much more than we need, and it's our fault. We are putting it there. In climate lingo, it's called a human-made "forcing." And it is that. We're not talking about a natural effect anymore. There's no getting around it. Someone else can explain that better than me. Too much carbon dioxide means too much absorption and downward emission, the Earth heating up faster than it can cool itself off. We are no longer in energy balance.

Will we one day, in violation of this fundamental principle, no longer be able to support our own lives?

Twelve Things Going through My Mind during an Average Minute of My Life on Retreat in Ojai

1. What is this life?

2. How to handle impermanence.

3. I just watched a branch fall from a tree. Another branch caught it before it reached the ground. It is held, suspended, vertical, delicate, and thin with yellow-green leaves.

4. I miss my husband and daughter.

5. I love this time alone.

6. I hear the heater on to my left, its periodic whir like a gentle dragon's breath in the room.

7. I love standing underneath heat pumping in, down, filling a room with warmth.

8. The irony of my love for ice and snow.

9. How much longer can I dodge COVID?

10. I think we all got it early, before they knew what it was. Steven's pneumonia. The ache of breathing as my chest expanded against water in the bathtub.

11. The quote at the beginning of *The Hot Zone*. Will the virus kill us all?

12. The practice of this life—to not be overwhelmed by how or when it will end for me, for them. To be content to watch that branch fall.

Destiny

I always believed that I was immortal. It wasn't that I thought I was a superhero. I simply couldn't fathom no longer existing, no longer breathing, thinking, being consciously aware. Falling in love, like *really* in love, made me mortal. Suddenly I had something to lose, and it made me somehow weaker, like Kryptonite. I knew love was strong, made you stronger, like you could do anything if you knew somebody would love you, whether you succeeded or failed. You could always come home to that person and be loved. This was powerful, and I felt that too. But I also felt incredibly vulnerable and tender, like the underbelly of a cat—that soft, long, rarely touched fur that hardly ever sees the Sun.

I felt that vulnerable again when some routine bloodwork that was part of an annual physical came back not so routine, not so normal. My breath caught, my stomach fluttered. I felt slightly off balance, leaning a little to the side like the Tower of Pisa. What did this mean? Was I going to be okay? More lab tests followed, and more, and more after that. I was stuck more times, and for longer each time, than I had been when I was pregnant. Fifteen vials of blood removed from my body on one occasion. I floated out of the medical building.

Some tests came back fine. Some were abnormal. What the hell was going on? An ultrasound of my liver came back normal. An incidental finding was a cyst on my kidney. More tests, more exams were ordered. The full workup had its advantages. Perhaps something asymptomatic that wouldn't have otherwise been found for years would be uncovered and addressed before it progressed to the point of no return. It also had its disadvantages. Checking for everything means that something will be found. And that thing must be followed up on. My medical team was doing their due diligence. However, in the process of all of these discoveries and flags leading to more tests and flags and on down the chain, my mind went into overdrive. Thinking turned into worrying and imagining the worst.

My fear had filled the whole world before. When my daughter's head size had continued to increase beyond what was normal for her age, the growth chart curve getting steeper with each doctor's visit, I didn't know how to stop being so afraid. It was as if I was in a rowboat heading toward the drop-off of a waterfall. I knew there were choices—jump out of the boat and swim to shore; stop rowing the boat, at the very least. But instead, I kept rowing, kept heading for the drop-off, even as I heard the long, deafening gush of water ahead. I just couldn't stop pushing toward the edge of all things.

Eventually it was surmised that she had simply inherited her large head from us. Finally, the rate of increase began to flatten out, allowing the rest of her body to catch up. Nothing to worry about in that department any longer.

Now I'd found something else to take its place. Would they find, when all the results were back and all appointments attended, that I had some terminal disease? Playtime with my daughter began to become poignant, wistful—my fearing not being here, and what that would mean for her, for my husband. We knew of an old friend who'd lost his wife young, forced to carry on and raise their children alone in the shadow of her memory.

How could this be a possibility for me? *Me?* Who had been a good person all my life and worked hard and done well? The utter arrogance. The ego. I wasn't used to being the one to whom loved ones expressed sorrow that I had to deal with a major health challenge. I got things done. If there was a problem, I solved it. I was my father's daughter.

When the fifth call in two days came, to schedule yet another test, I lost it. I had nowhere else to turn. So I turned to God. That's what I call my higher power.

I'm not a religious person, but I am a spiritual one. I must concede that there are things in the universe, even on our back patio, that cannot be explained. I couldn't think myself out of my fear of falling forever into oblivion, never to be seen or heard from again, never to see my daughter or husband again. My mind had reached the limit of its capabilities. Between that limit and the transcendence of a temporary reprieve from my thoughts of doom—that was where God needed to step in. So I prayed.

I got down on my knees and told God that this all felt like too much. Too much to deal with along with a full-time job and a toddler. Too much to deal with, period. I told God that I was scared. And then I asked for help.

My phone buzzed. A friend was asking me to come speak. Someone had canceled at the last minute. Could I fill in? I smiled. *Ah, God, so this is what you have in mind. Get out of your own head, Aomawa. Help someone else.* Message received.

So I did. And for that hour, I didn't think about my fear of dying before I'm ready. The truth is, I'll never be ready. And being healthy is no guarantee of my longevity, if we're being honest. Everyone knows someone whose life ended all too soon through outside forces of destruction rather than disease and biology.

It turned out that I'd gotten a virus akin to mono that had done a number on some of my organ enzymes. I remember feeling lousy for a couple of days, and then I felt fine. It had taken much longer for the bloodwork to say the same. Thankfully, the bulk of it all was temporary and soon returned to normal. The only thing I would need to address was my thyroid, which sometime after giving birth had decided to become an overachiever. I am sure I have *no* idea where it got that idea into its head.

What I have learned from this recent experience is to bring God with me through it all. I don't know why this was such a revelation to me. I've seen God work in my career. I thought I had to choose either science or acting. It couldn't be both. There was no way I could combine both. How would that even be possible? Then the opportunity to host a science TV show for PBS fell into my lap. What I'd neglected, what I was reminded of later by old friends, is that God is not limited by my lack of imagination. There were options, possibilities that I simply hadn't seen. In that space between what I thought was possible and what actually was—that was where God hung out. And none of my faith needed to conflict with my ethics or principles as a scientist. Someone whose name I can't remember once said that science and religion (I'd say "spirituality" here for its broader and more inclusive aspects) should never be in conflict. If they are, then one or the other is overstepping its bounds. Carl Sagan wrote, "Science is not only compatible with spirituality; it is a profound source of spirituality. . . . The notion that science and spirituality are somehow mutually exclusive does a disservice to both."

Science and art too are not so disparate as some might argue. They not only complement each other, they overlap. They have in common the seeking of a profound spiritual appreciation for the creativity that permeates all aspects of this incredible world and universe.

I've heard that I can choose to believe that God does not exist, and look for evidence to support that, or I can choose to believe that God does exist, and look for evidence to support that. Either way, I will find the evidence I am looking for. I choose to believe that God does exist, because in the act of searching for the evidence to support my theory, not only do I find it, but I find something else along the way that's perhaps even more important: humility. I see myself in my rightful place compared to others and to something much, much bigger than me. I am not God. I have acted like I am from time to time, trying to control things over which I am powerless, thinking I can predict the future or know what is best for others, or even for myself. I have been shown, over and over, that I really don't. This is why I will never tell a student that they shouldn't be a _____, whether that's an astronomer or a pianist or an airplane pilot. I'm not that person's higher power. I have no idea what they should become. I can tell them that they need to work a lot harder to do well in my course if they're struggling. I can ask them, "How badly do you want it—this career? How much do you love it?" A dear teacher once told me that if you love something, you'll be good at it. But I don't think that's necessarily true. If you love something, it means you'll be willing to work as hard as you have to to *get* good at it. That's why I ask students how much they love the thing they want to do. If it's a lot, then they'll work and work at it, and eventually they'll get better. They may even get really good.

The only amount of time I can depend on as a scientist—the time for which there is a probability of 1—is this present moment, today. I lead students in all my classes and team meetings in a few minutes of secular, mindfulness-based meditation practice. I encourage them to focus on their breath, each breath, in and out, over and over again. Our minds swirl in the space above our heads, and that's normal. We keep bringing our minds back to our breath, to our feet firmly planted on the floor. To this room, this place, this time. Not yesterday, not tomorrow, not an hour ago. Right now.

I do this for them, to help them learn to stay in today, to remind them that it is all they have. I also do this for myself. This moment, my daughter's eyes bright and sparkling blue looking into mine, is all I have, and all I need. I am so glad, in all three times applying to the astronaut candidate program, that I was never selected. As much as I love and am inspired and comforted by space and discovery, the dream of setting foot on a new world, it pales in comparison to the thought of being away from my family for the time it would take to get there. We must have priorities, or we will lose everything important amid the dark churning sea full of what isn't.

When my daughter says to me, "Come over Mommy, play with city," beckoning me over to the miniature collection of roads, cars, and helicopters in front of her, it is the most precious and important moment of my life. It matters so much more than the small stuff, the fears and embarrassments, the neuroses and anxieties. The "not enough" and the "should." It is the bright red rose in a great wide open field of grain. Walk to the rose, bend down and sit in front of it, and breathe in. Will you breathe with me? Close your eyes. Take a deep breath in through the nose. Now sigh it out through the mouth. Again, in through the nose. And sigh it out through the mouth. Wiggle your fingers and toes. And when you're ready, you can open your eyes.

☽

I had lunch with Ann Druyan, Carl Sagan's widow and writing partner. She had emailed me out of the blue a couple of weeks earlier. Her son, Sam, had sent her my TED Talk. She hoped to meet me and asked if

I would come to have lunch with her in LA, where she was in post-production for the new season of *Cosmos*.

My mouth hung open as I read the email. *Carl Sagan's widow wanted to meet me.* The coauthor of *Contact*, one of my favorite books, wanted to have lunch. The day I found out that I'd been accepted into graduate school in astronomy the second time around—the time when I finally, ultimately, and wholeheartedly wanted it badly, this life in astronomy—the first thing I did after I hung up the phone was look up at that poster of Carl Sagan sitting on a sea of new, uncharted worlds waiting to be explored.

Now I was meeting the most important person in his life.

I sat down at her cozy table at the posh Belvedere restaurant in the Peninsula Beverly Hills hotel. We talked over salmon and potatoes. So many stories of Ann and Carl. How they met. How they got together. The day they found out about Carl's illness. I sat in rapt attention and gratitude for being trusted with these special moments in their lives. I was present—fully present—and at the same time, I felt like I was oddly hovering above us both, circling the rafters of the restaurant like a cartoon ghost looking down at it all.

I ordered less than I wanted to eat. I was still nursing, and ravenous all the time, but we didn't know each other well yet, and I knew she would be paying for lunch. I hit a drive-thru on the way home, stuffing my face with a hot fudge sundae and fries.

As we said goodbye, she clasped my hands and told me how wonderful it was to meet me, and she meant it. Our lunch had told me one thing about her—she meant what she said, or she didn't say it. She would prove it a year later in her toast to my tenure from Ithaca, New York, via Zoom during the pandemic. I sat there again, hovering outside of my body as I listened, and hoped Carl would have been just as proud.

Months passed. Ann and I exchanged warm emails. She took a vacation. My life got busier. I was offered a stint at the Kavli Institute for Theoretical Physics in Santa Barbara, and Steven, Garland-Rose, and I went up there for a month so I could work on my energy budgets paper without teaching responsibilities. I broke my foot. It healed. School started. The tenure dossier loomed. Teaching. Papers. Mentoring of students.

But just before a global pandemic caused the world to stop, *COSMOS: Possible Worlds* premiered at UCLA's Royce Hall. It was my first big night out since before Garland-Rose was born. I'd wanted all three of us to go, but the more Steven and I discussed it, the less feasible it seemed to cart a toddler to a late-night red carpet premiere. So they both stayed home. I sat next to Bill Nye on my left, my friend Denise, whom I'd brought with me, on my right, and Ann and Neil and the rest of the heads of production up onstage talking after we'd been shown episode one. At the party afterward, between passed hors d'oeuvres and planet-themed cake pops, Sam Sagan found me and carved a path through the crowd surrounding Ann. We shared another warm hug. Then I sped away back home to Irvine.

There is a genetic principle called convergent evolution in which two unrelated species evolve to exhibit a similar trait. Insects, birds, and bats, for example—all unrelated species, all possessing the similar biological machinery of flight as a result of the need to adapt to similar environments or ecological conditions.

Carl Sagan and I had come from different worlds on planet Earth. Carl was a gifted scientist with a deep commitment to understanding the wonders of the natural world using the fundamentals of the scientific method—hypothesis, experiment, analysis, and interpretation of results. Though he was from a Jewish family, his "religion" was science, and while his explanations of the workings of the universe were secular, he found science to be, in his words, "a profound source of spirituality." I interpret this to mean that when he contemplated the vastness and scale of the universe—the many processes and perfect elements aligning to create a planet, its oceans, the life swimming within, becoming you and me walking around and breathing, all of our trillions of cells allowing our function, our ability to love and express our love to others—it was enough to provide the same feeling of rapture, awe, and humility that others feel sitting in a church pew staring at the illuminated stained glass while a sermon rains down, manifesting the presence of God in the room.

Carl never saw a contradiction between science and spirituality. And neither do I. I will, however, go one step further than Carl might have, and say that I believe in both science as a profound source of spirituality, and in God as my ultimate source. I believe there is no one without the other. As much as we have been able to explain through the application of the scientific method, in the lab, at the telescope, in the deepest ocean, in the driest canyons, and at the highest mountaintops, we are left with many orders of magnitude more that remains unexplained. We know that in the tiniest fraction of a second after the Big Bang (as in, a few trillionths of a second), an incredible burst of expansion occurred, creating the makings of everything we know and hold dear today from a dense nothingness. But what caused the Big Bang? In that space, in the unanswered question—that is where a power beyond all reasoning and scientific interpretation resides. That higher power lives in miracles, and in the breath between understanding and everything that is still not understood. I can say this now without fear of being called a crackpot.

I practice the discipline of science, and I will continue to practice science and use it to explore and interpret the worlds around me. Science—specifically the rock record—tells me that the Earth is 4.5 billion years old. Science tells me that the likeliest origin of the first life on this planet was a warm, wet environment, such as a hot spring. Science tells me that around nearly every star there orbits a planet. My intuition tells me that with 10^{22} stars in the observable universe, and at least one planet around almost every one of them, it is unlikely that ours is the only planet that hosts life. Intuition is not fact. To turn intuition into a fact that science tells me, I need evidence.

Science has told and taught me a great deal. And yet, science can't explain everything. I carry the coat of science I wear lightly, with an inner lining of humility, as I can see that there is a limit to what even the greatest scientific minds of human history will ever know or understand. I am not one of those great scientific minds, like Carl Sagan or Albert Einstein or Galileo Galilei or Isaac Newton or Marie Curie or Katherine Johnson or Rosalind Franklin or Vera Rubin or Gladys West (and if any of those names are unfamiliar to you, go look them up now). I admit to this not because I still live in my impostor thoughts, but because that isn't my ultimate path. I am someone with the arts in my blood and a strong drive and ability to apply myself, powered by a deep wonder and love for the universe that gave rise to the desire to dedicate myself to its exploration and to sharing it with others. That is the evolutionary point at which Carl's and my divergent paths converged. We both love sharing the wonders of the universe with the world. Carl continues to do this through Ann, through Neil, through recordings and books. Death is not the end. Whether there is a heaven or not, love continues on.

I myself have had too many stars align throughout my life to continue to chalk them up to coincidence. I know a divine power beyond my understanding is at work in my life. I believe one is at work in your life too. If that doesn't sit well with you, then I hope you'll forgive me for saying it.

I do love the mystical, and I enjoy taking inspiration from many sources, from my horoscope to tarot cards, and people like Louise Hay, Wayne Dyer, Deepak Chopra, and Thich Nhat Hanh. I don't call any of that science. But who says inspiration must come from a single spring? That's how a river runs dry. How plentiful the spring that is fed from multiple channels. And how abundant.

Home 2.0

The city where you are wakes. The din rises up into the early morning haze. Or if there is no din, if you're in the countryside, somewhere quiet, still there's the Sun's pale egg emerging out of it all to oversee, each soft ray pulling something out of the fray, illuminating a story. What is your story? Yours, right now, in this moment in time. Where is home for you?

For me, home is comfort, a cushiony chair, my bed covered in blankets and pillows. Home is our living room, where most of the life is, our television, where our daughter watches *Moana* and *Tumble Leaf* and *Frozen II*, over and over again; where our dining room connects to an open-air kitchen, and all of the food we eat is prepared, then placed on the blond wood rectangular table, and she climbs up into her chair, seated on a large *Marvel: The Avengers* coffee-table book to prop her up so she can reach her plate of turkey, carrots, and apples.

Home is my office now, in a world filled with disease and a slow-creeping vaccine. I teach at home, I write at home, in the corner room of the house, upstairs, in my home office full of books on exoplanets and planetary astrobiology, standing up next to books on not being such a scientist when you communicate, next to poems by Rita Dove and

Denise Levertov and *A Challenge for the Actor* by Uta Hagen. There is no separation now.

In the aftermath of the tenure quest, I have rediscovered who I am. That professional development program for faculty that I'd discovered as a postdoc, NCFDD, helped me find it. Tara Mohr's *Playing Big*, which I stumbled upon in Bart's Books, the famous outdoor bookshop in Ojai, helped me claim it. I am a champion of interdisciplinarity. Kindness, spirituality, and healthy personal–family–work life balance are my core values. I love to dance. I love to write. I do yoga. I meditate. I read poems and memoirs. I love beautiful cards with affirmations, the ocean, and a good leather purse. I dream of living in Italy and Ojai with my husband and daughter. I paint, I stargaze, and I take care of my skin. I share my feelings, and I try not to make them someone else's fault. I love putting money into savings, and I don't use credit cards. I believe that needs are important, and wants are too, and that there is enough in the universe for both. My favorite word is *fiasco*, and *shenanigans* is a close second. I play the violin intermittently. I connect with the Moon. I have a strong, clear voice. I protect my time. I love snow and ice, and I get cold in 65-degree weather (see the both/and at work here?). I get sayings mixed up so often that in my family we call them "Aomawa-isms." I sometimes get confused about when to be a disciplining mom and when to be a play-like-a-kid mom. I hold grudges sometimes. I say sorry when I need to. I let my daughter have her feelings. I let myself have my feelings. I practice basking in my own glory, and worry less about whether someone thinks I'm pretentious because of it. A personal goal is to be unabashedly bigger than my britches. Every year I grow. I need at least eight hours of sleep a night, preferably nine. I prefer Double Quilted Northern toilet paper to any other kind. I drink tea, not coffee, and genmaicha tea is my favorite. I love writing down things that inspire me in notebooks. I like to sleep in, and at the same time I love seeing Garland-Rose open our

HOME 2.0

bedroom door and come in at 7:15 with her Maui and Te Fiti figurines and place them on the corners of Steve's and my bedside tables, respectively. I have an overactive thyroid that is calming down on medication. I love reading to and being called "Mommy" by my daughter. I'd like to watch a glacier from a cruise ship in Alaska. But I wouldn't want to see it fall into the ocean, because that would make me think about the negative impact of climate change, and sometimes it's nice to take a break and pretend that the world is okay. I'd love to find out how your personal journey shaped who you are in your work. I also love rings. I love my Fitbit. I like to hike. I'd like to camp more, someplace safe. I could get by on a desert island with soy milk, lip balm, hand moisturizer, something spicy, like Jamaican beef patties, and almond macaroons from Urban Plates. I say "no thank you" a lot at work. I still sometimes take on too much. I am growing braver. I am protective and loyal. I try to empower my daughter. I do things for ten minutes. I truly believe it's never too late. I buy books I can hold in my hands (but if you're reading this on a Kindle or an iPad or your computer, thank you—you are helping the Earth). I love the smell of lavender. I love massages, spas, and all things related to pampering. I love being enfolded in my husband's big, strong arms. I was surprised to discover after gaining tenure that I like being a scholar, not because it gets me promotions, but because I like discovering new things, sharing my expertise, and interacting with colleagues. I am firm when I need to be. I get antsy for travel sometimes. I love coming home to my family who knows and loves me exactly as I am.

    I do all of these things now, and am all of these things now. It is all melded—the worlds, the universe outside with the universe within. I eat, sleep, shit, fuck, sing, read and in the same home I teach, speak, write, grade, recite, instruct, cry, shout, apologize, scream, and dance my ass off. I danced this morning, like I danced on the last day of the quarter, instead of tuning in to another Zoom meeting. I turned off the lights in my office and turned up the volume on my computer, closed my eyes, and started moving. My body was an arch, a bridge, it was a valley, then a cavern, I crossed a chasm and strung my legs between the ends of the rope, linking the sides. The rest of my body walked across and folded in on itself, then shaking, hard, knees up to my chin, running, then stopping to jump, then stretching to my left, reaching my left arm as far as it would take me, then a rubber band back, then the other side, and out as far as I could go. The band Fenech-Soler was with me, the pulsing, and with it my memories of the '80s and being free, pining for some boy, imagining the end of the Brat Pack movie, except I was Molly Ringwald and he was Andrew McCarthy and there was no one else in the world but us in the middle of a dark, wet street. I was in it now, and it was all there was—my body, the music, the best moments of my life, the ones I dreamed would happen, and everything in between. The walls were covered with pictures of my visions realized—the trip to Italy, the couch, car, home, bed, child. I opened my eyes and looked around as I danced, and I saw it. And for a fleeting moment I thought that maybe I didn't need any more dreams.

Journey

These ices. They are how I started in this field—falling in love with a property of water ice that was like me. I couldn't stop there. Pure water ice led to salty ice, then to land. And now, after we've studied all of those surfaces, there are so many more. So many forms of ice that can exist when it gets cold. And so many planets hurtling through space in strange orbits that take them searingly close to and then frigidly far, far away from their stars. The atmosphere bows to these whims, pouring out onto the surface as solid ice, and then sublimating back up into the sky. How does this process affect life that might exist on these worlds? How does starlight interact with these exotic ices that were once gases that made up the very air above, that life might have breathed? And can that ice become that air again? Over how much of the orbit that comprises the planet's year? Models aren't ready to answer these questions. And they need to be. We will get them ready so that they can.

It was fitting, in the end, to have worked on the proposal, though I hadn't seen it that way on the day I left. This proposal was my grad student's, to support her dissertation. But my name was on it, so I had to do a lot of work to get it into respectable shape to be submitted. It was the kind of thing my PhD advisors had done for me all those years ago. Now it was my turn. I hadn't signed up to work on this thing to begin with. My student wanted to apply for it. What was I going to do, tell her no? Leave her out to dry with no guidance? I held off for as long as I could, to give her a taste of the degree of work involved in trying to convince a group of people outside of your field that what you want to do is important; so important that they should throw hundreds of thousands of dollars at you so that you can do it. It's a grand task. I've done it successfully four times—once as a grad student, once as a postdoc, and so far twice as a faculty member. Now I teach grad students how to do it in a course on communications in physics and astronomy. As much as there is to learn, it isn't an exact science. I've had proposals rejected nearly as many times as they've been selected. Sometimes luck is on my side. I didn't know it then, but the extra work would pay off. This proposal of my grad student's was selected for funding.

After the inevitable last-minute issues caught by conscientious administrative personnel, the thing was at last submitted, and I backed my car out of the driveway and started off, three hours later than I intended to. My first trip alone since before my daughter was born. Back to Ojai, to write and rest. I smiled as my navigator app took me along the Pacific Coast Highway for more than half of the drive. The ocean on my left, the Sun sinking low. I felt my shoulders relax, my hips sink deep into my seat. There were shadows in the water—surfers in sleek wetsuits waiting for the waves with their names on them to crest. Six seagulls hovered above the water in formation, then tilted in perfect alignment, a straight line rocking diagonally across the sky. I passed the old beach Thai restaurant I used to escape to from LA when I needed a day off. I passed that dive for the motorcycle crowd where Steven and I got food poisoning. I passed Zuma Beach—another escape when I needed to get even farther away from civilization. The highway snaked between high cliffs with signs on the road that read WATCH FOR FALLING ROCK, and cars parked to watch the sunset, some people picnicking on the edge of the rocks, peering over into forever as the ocean nearly rose up to meet them. All of this, even in the middle of a pandemic, could not be missed. It was life, it was the wonders of the Earth orbiting the setting Sun, and it all happened every day. That day, I had not missed it.

I rushed to pick up dinner at a Lebanese rotisserie chicken restaurant in Ojai that I'd liked the last couple of times we'd been there. They brought it out to my car. I texted Steve that I'd be ready to talk with him and Garland-Rose on the computer for dinner in about thirty minutes, and I raced to the grocery store. Sucked into the giddiness of shopping for just myself, I threw four different kinds of cheeses into my cart along with French bread, fig-orange spread, parmesan crisps, fennel, cherry tomatoes, basil, apples, bottled health food smoothies, immunity and wellness shooter drinks, antibacterial wipes, and wine. I hadn't done this—stood in front of a wall of bottles reading the descriptions on the back, looking for the words *supple*, *soft*, and *low tannins*—in years. Nursing over, I was free to consume whatever I wanted without thought of who else was going to be consuming it next.

I was late checking out, and Steve and Garland-Rose were waiting. As I stumbled into my room at the Ojai Retreat in the dark and fell into a chair to get my laptop set up, my car full of groceries and luggage and yoga mat, paint supplies, bags of books I might want to read during my stay, books to read to Garland-Rose during our daily calls, I felt frazzled. Garland-Rose didn't notice or care. She wanted a book read. My dinner waiting in the bag next to me, I started reading. I watched a smile spread itself wide across her face to see me, to hear me reading one of her favorite books. She turned to Steven with her big grin. He reached out and held her hand as they listened. Even though I wasn't in the room with her, this was enough to calm her small, warm body. I felt relieved. The plan had worked.

Three books read, we air kissed goodnight and hung up. I looked around the room. I remembered it. The one with the full kitchen (good, because I bought enough groceries to fill it). The one with my favorite balcony, enclosed, looking out at a canopy of trees and the towering mountains above the valley below. I tore pieces of roasted chicken from the breast bone with my hands, dipped it in garlic sauce, and ate it ferociously, along with roasted potatoes and spiced bread dipped in thick yogurt that was mixed with charred lemons. Who needs utensils? I licked the juices from my fingertips.

After three trips back to the car to unload, I unpacked while watching *Eat Pray Love* on my phone. It felt appropriate. Elizabeth Gilbert had journeyed to Italy, India, and Indonesia to find something—herself. I hadn't traveled as far—just a two-and-a-half-hour drive. And I was pretty sure that I finally knew who I was. But amid all of the responsibilities of life—work, family, errands—the volume of who I was had gotten turned way down, while everything else was turned way up. I was fighting for time for me, time for this book, time to just sit still and do absolutely nothing. I didn't need to bring half the stuff I brought with me. But when you're faced with two solid days to yourself—no demands on your time, no responsibilities, no schedule—it's like being let loose in a candy store with no spending limit. I thought I might want to do it all—everything I hadn't had time to do for the last three years. I'm a Girl Scout, or a Girl Guide as they call it in Canada. I was prepared.

It took time to settle in—rather, to settle out of "doing" mode. I knew enough not to check my work email, and I had my away-from-email message on that account. I opened my personal account and saw an email from an old Exeter classmate. We had just lost a classmate to suicide, and it had hit everyone hard. People reached out to each other in the fog. She was writing on behalf of a friend of hers, whose daughter was in love with astronomy and looking for a way in. I leapt in, eager to get this request off my plate so that I could get back inside my retreat bubble, secretly cursing myself for even reading my personal email. Two additional follow-up emails later, full of suggestions for this girl on the other side of the country whom I'd never met, I deleted the entire email chain, set down my phone, put a spoonful of granola with black-cherry yogurt into my mouth, and opened my eyes wide. The mountains in front of me rushed in.

I want to say that I'm cured. I completely believe I'm entitled to be a scientist and an astronomer, and I have it going on. I want to say that it feels natural now, like a second skin, and I don't ever think about sounding smart and having to prove something. I want to say that I finally realized, as I stood explaining a plot in group meeting, that this is the only thing I could ever do, that I was born to do this. This is my passion, my blood. I want to say all of that, and mean it.

But there will always be doubt. I will never be able to say, "This is the only thing I could ever do, ever wanted to do." But this was the first thing that I wanted to do that never truly went away, even when I went away from it—far away, to where I thought I'd hidden from it behind glamour-laden nights and store shelves. It found me still. I want to say that it will get easier, the reading and the writing and the talks and the sharing of ideas and thinking of new ones, and writing papers and having to be learned.

Sometimes, I think that life was easier when I was a struggling actor. In Hollywood, aspiring Black actresses are a dime a dozen, not an endangered species. Hollywood still has a long way to go. But not as far as academia. Not as far as the sciences. Not as far as physics and astronomy. There's a reason this world is called the Ivory Tower.

I can't do this for them, for Black women everywhere. That's not enough. It won't keep me warm. It won't fire up my belly with satisfaction and inner glory—what I needed to get through graduate school, what

I needed even more to get tenure. And for what? More Black women everywhere, so there can be more than five of us with faculty positions in this country? That's not enough. For the villa in Tuscany that Steve and I and Garland-Rose will get on my sabbatical? Close. But not enough.

It's time to look at the Moon again. It never fails. It's a big scary thing, growing. Getting bigger, life expanding. The limitations on paper slip away, and I am left with the memories of an MIT student walking across the bridge one Saturday night to buy a bottle of wine and a single wineglass from Crate & Barrel because she'd had too much and it was all too much.

Many years later, I crossed the same bridge, my black TED backpack on, my dear leather purse across my body, a MacBook Air full of my work, my data, and a whole full life behind and ahead of me. To make. To break. To grow. For me.

A blue jay lands on the stone wall on the balcony outside my window. She is brilliant, and a few seconds later, as she disappears into an adjacent tree, I hear her calling to another bird. It is a loud conversation. Other birds swoop into the branches. A meeting is convening. Above my head two hawks are circling each other over something far below. I watch their wings spread wide and almost still, gliding on a waft of breeze. Thousands of feet above them, a bright white airplane crosses overhead. An incredible feat of innovation for us to be able to fly, like these birds. Still, the grace of the hawk is incomparable.

I look out at the tall mountains awash in sunlight, tracing an uneven line along the edge of the milky-blue sky. They would wait patiently for me to be ready, to be still too, to breathe deep and trust that I am whole and enough, just as I am. They always would. Maybe I could be patient with myself like those mountains. Being gentle never hurt.

Harder still than Hollywood, than academia, is a life strung somewhere in the middle. The in-between spaces, not quite fitting in anywhere, is a difficult road. It was made easier though, or at least not as hard, when I let go of trying to fit in. In the poem by Mary Oliver, she says it. I knew what I had to do. When I embraced what I had been running away from—that I was indeed different, and that there was nothing whatsoever wrong with that—I was at last set free. I had saved my own life.

The trick is to stay free, to keep saving my own life, over and over, for eternity. To work at being comfortable with choosing, then choosing again, and again, then finding, at last, the perfect combination of avenues and interests and outlets to feel yourself 100 percent represented and expressed. Even then, there will be times when you doubt, and fail, and sidestep, and forget, and then remember, and fall, and then get up and step forward. This is what it is to be human. Unfortunately, what is also human nature is judging ourselves in these moments. We think we should be past it all by now.

I watched my daughter stumble over a bush the other day because she was looking up at the sky while she walked. That's my girl. I see her discover and explore our world as I once did. As I find new planets to explore beyond our solar system, I am reminded that there are no boundaries to what is possible. I am simultaneously held firmly and gratefully to this Earth by the immense gravity of the new worlds my daughter and I are discovering together, full of hyphens and dashes and *and*s and whole, complete lives.

What would it be like to accept that as we keep growing, we keep practicing and learning how to accept that we are an evolving species, on an evolving planet, inside an evolving and expanding universe? We aren't meant to be static. We aren't meant to "get it" and be done. We are works in progress. And we never know what's around the corner waiting to help us grow in our development as human beings.

I want so much for you. You may not believe me. How could that be true when I don't even know who you are? It's true that I don't know you. And it's also true that I want so much for you. If you haven't learned anything else after reading all of this, surely you must know by now that both things can be true. Both can exist. Both, and more.

What do I want for you? I want you to look up and be amazed. I want you to feel supported, less lonely and afraid, a part of rather than apart from. I want you to ask that question you have been wanting to ask, and let go of worrying whether someone already asked it while you were daydreaming, or whether someone will think it's stupid or impossible to answer. I want you to know, but not just know, *feel*, deep down in your belly, that who you are is magnificent. I want you not to merely tolerate, but to celebrate your many contradictions, and embrace the full, complex being that you are. When you look up at the sky, clear or not, don't you know that there are hundreds of billions of stars looking back at you? You are on a planet rotating on its axis at a thousand miles an hour and hurtling through space around its Sun at nearly seventy thousand miles per hour, as the Sun rotates around the center of our Milky Way galaxy at over four hundred thousand miles per hour, while the galaxy moves through space at more than one million miles per hour. And this same thing is happening on hundreds of billions of planets orbiting hundreds of billions of stars moving in hundreds of billions of other galaxies elsewhere in the universe. So much is happening right at this very moment, as you breathe this breath.

This needn't make you afraid. What it can do, if you let it, is make you simultaneously humble and expansive. There is so much beyond your own individual struggles. And at the same time, what you want—everything that you want—the universe can hold it all. Truly, it can. Looking outward can be a good thing, if you know where to look. Look up.

In the past, the field of astronomy assumed life would be found around stars that looked like our own Sun. Today, the field is recognizing that planets with the conditions for life might—and perhaps are—most likely to be found around unfamiliar stars—those red ones I've told you so much about, that I've spent most of my career thinking about, imagining how the grass might feel under my feet on a planet with a red Sun overhead. Over the next ten years, the field will identify thousands of new planets around these stars, and many of those may be good candidates for life-bearing worlds. How surprised will our species be when one day we discover the nature of that life elsewhere? How likely is it to be like anything we think? Will we admonish ourselves for not anticipating its form, its biological structure, its modes of communication? Or will we simply allow ourselves to be surprised, and revel in the joy of discovery? Could we treat ourselves like that now? Like the single greatest discovery of our lives?

ACKNOWLEDGMENTS

Here I will attempt to thank all who made what you just read possible. As you'll see, it takes more than a village to raise a book. In my case, it took a planet.

Thank you to my agent, John Maas, who, long before all of the reading and guidance, reached out across time and space to ask the question, "Would you like to write a book?" That single "Big Bang" moment helped to create everything that came after, and ultimately, what lies before you. Thank you to Celeste Fine for sharing wisdom, knowledge, and celebratory toasts, and Mia Vitale, Susana Alvarez, Anna Petkovich, Elizabeth Pratt, and everyone at Park & Fine Literary and Media.

I had amazing editors at Viking throughout the writing of this book. It started with Georgia Bodnar, who in all of our communications stayed true to the shared vision that even with a book about a scientist, the personal narrative drives the story. Gretchen Schmidt gave incredible feedback on an early draft of the book. Emily Wunderlich, my final editor, carried us—book and me—across the finish line. Thank you for helping me to find the clearest way to tell this story. Thank you also to Emily's assistant, Paloma Ruiz.

Thank you Andrea Schulz and everyone at Viking and Penguin Random House for your enthusiastic support of this memoir and its author from day one.

Thank you to my assistant, Christina Dinh, for helping me manage all things related to being a professor and a book author while maintaining balance in life.

Thank you Emily Margolis, Sara Schechner, Mary Dussault, David Jones, and the Department of the History of Science at Harvard University for providing valuable insight into the history of women at Harvard-Radcliffe.

ACKNOWLEDGMENTS

Thank you to my lawyer, James Gregorio, and my accountant, Patrick Tuttle, for their thorough attention to contractual and financial aspects related to the book.

Thank you to my parents, Margaux Simmons and Idris Ackamoor, for the gifts of life and love, and for my creative, strong spirit.

Thank you to my brother, John-Samuel MacKay, and sister, Kayla Edwards, for their love and support, and to my stepfathers over the years, John MacKay and Kirk Edwards, and my mother-in-law and father-in-law, Sharon and Gary Shields, for loving me as their own child. Thank you to my sisters-in-law, Lori, Sue, and Jann, and brothers-in-law, Stephen, Bob, and Anthony, for welcoming me so warmly into the family. Thank you to my mother's partner, John Lassell, for joining our family.

Thank you to my writing teacher, Natalie Goldberg, for teaching me how to write and to practice living as an artist and a human being in the world.

Thank you Susan Allison for guidance, and unconditional love and support.

Thank you Allen Zadoff for sharing experience and providing guidance throughout the making of this book, and long before that.

Thank you Silas Munro, Lara Hoefs, Alwyn Wright, Hayley Buchbinder, Lisa Paris Repice, Sand Chang, Liye Xie, Maggie Tennesen, Terrell Lozada, and Sue Ann Pien, for time, and strong encouragement and support in helping me take care of myself during this process.

Thank you to my friends for consistent love and support through the phases of single gal, married gal, and married-gal-with-kid-and-two-other-jobs—Yvonne Lopez, Ashley Barrett, Denise Tarr, Kara Revel, Jonna Hackett, Shellyann Fluker, Nina Irani, Catherine Shiao, David and Catherine Chittick, and Alex Lee.

Thank you to my academic coach, Mary McKinney of Successful Academic, for counseling me on work and life as a pregnant professor and then working mother on the tenure track, throughout the transition to tenured faculty, and into this next phase of professor life off the beaten path.

Thank you to my therapist, Jessica Hadlock, for helping me to recognize the power I have to replace old narratives that no longer work for me as a mom and a human being with new ones that help me experience joy, peace, and ease.

Thank you to our couples therapist, Bert Shepley, for help rediscovering the "we" after many years of marriage and then becoming parents.

ACKNOWLEDGMENTS

Thank you to Sarah Stewart Johnson and Daniel Whiteson for sharing their experience with writing books for the public while balancing jobs as faculty in the physical sciences.

Thank you Genevieve Scott for sharing experience about the writing, editing, and publishing process while our kids rambled around during playdates and birthday parties.

Thank you to my cousin LaJoyce Brookshire for valuable conversations on book publishing and promotion.

Thank you to the National Center for Faculty Development and Diversity (NCFDD), especially its Faculty Success and Post-tenure Pathfinders Programs.

Thank you Tara Mohr, Rosemarie Roberts, and Michelle Boyd at Inkwell Writing Retreats for guidance in taking big leaps.

Thank you Jedidah Isler, Lia Corrales, Camille Avestruz, and Catherine Espaillat, founder of the League of Underrepresented Minoritized Astronomers (LUMA), for the consistent support of our postdoc (then faculty) women of color telecon over many years.

Thank you Tom Reilly and the TED Fellows Program for the opportunity to share my message on a global stage, directly leading to the opportunity to write this book. Additional thanks to Shoham Arad and Lily James Olds for help letting the world know about this book.

Thank you Chris Harper, Deidre Hunter, and Jim Elliot for supporting my dream of becoming an astronomer.

Thank you Tom Orth for giving me the opportunity to become a classically trained actor, and Mel Shapiro, Judy Moreland, Paul Wagar, Peter Wittrock, Meg Wilbur, Salome Jens, Jean-Louis Rodrigue, José Luis Valenzuela, Leon Katz, Gil Cates, Amen Santo, Jacques Heim, Andre Paradis, and Beverly Robinson for training me.

Thank you to John "Jay" Gallagher and Eric Wilcots at the University of Wisconsin–Madison for giving me a first chance to become a professional astronomer.

Thank you to Suzanne Hawley, Victoria Meadows, Cecilia Bitz, and the entire Department of Astronomy at the University of Washington for the second chance to become a professional astronomer. Thank you to my entire cohort in that graduate program for their support—Yusra AlSayyad, Amit Misra,

ACKNOWLEDGMENTS

James Davenport, Lori Beerman Payne, Yumi Choi, Patti Carroll, and Lauren Anderson.

Thank you to Kartik Sheth for sharing valuable tools that helped to set me up for success in my second PhD program, and for continuing to be a strong ally post–grad school.

Thank you to Minorities Striving and Pursuing Higher Degrees of Success in Earth System Science (MS PHD'S), its founder, Dr. Ashanti Johnson, and my mentor in that program, Dr. Akua Asa-Awuku.

Thank you Dr. Michael Kane and the process group for graduate women of color that he facilitated at UW.

Thank you Nancy Finelli and the UW Women's Re-entry Program for crucial resources and support.

Thank you John Johnson and Brad Hansen for mentorship, sponsorship, and encouragement.

Thank you Neil deGrasse Tyson and Claudia Alexander for guidance on establishing my "street cred" as a scientist before diving into the world outside of academia. It was hard to be patient at times, but I'm glad I listened.

Thank you to Ann Druyan for reaching out to me years ago to have lunch, signaling the beginnings of a relationship that I hold dear and look forward to continuing to nurture in the years to come.

Thank you to Tim Tait and James Bullock for being the chair and dean that they are, respectively, and supporting me professionally in the writing of this book.

Thank you to the University of California, Irvine, for the honor and privilege of being a professor at this institution. Thank you to the UC President's Postdoctoral Fellowship Program, Kim Adkinson, Mark Lawson, and Douglas Haynes for making connections, and for being long-standing champions.

Thank you to my incredible team at the Shields Center for Exoplanet Climate and Interdisciplinary Education (SCECIE), past and present—Ana Lobo, Vidya Venkatesan, Nicholas Duong, Kiana Whitfield, Maya Silverman, Jessica Nicole Howard, Eric Exley, Andrew Rushby, Igor Palubski, and Christina Dinh.

Thank you to the National Science Foundation for its support of the research and education projects described in this book (Award No. 1753373). Thank you also to the National Aeronautics and Space Administration for research funding (16-HW16_2-0003, 20-ASTRO20-0187, 22-XRP22_2-0189), and the Clare

ACKNOWLEDGMENTS

Boothe Luce Foundation for its holistic support of female faculty in the physical sciences.

Thank you to all the past, present, and future participants of Rising Stargirls programs for your commitment to exploring a new approach to learning about the universe. May you continue to invest in yourself as your journey of discovery continues.

Thank you to my daughter's school, UMS, and Cecelia McGregor, Alma Colin, Rita Piceno, Stacey Barry, Kelly Vu, Melanie Bragstad, Amanee Ayesh, and Amanda Cordova for teaching, childcare, parent's help, and date time with my husband. Thank you as well to Maria Navarro and Doradia Vazquez for loving care in keeping our house clean on a regular basis.

Several books comforted me like old friends while I wrote this one. In particular, I am grateful for *Mindful Moments for Busy Moms* by Sarah Rudell Beach, *Momtras* by Kristine McGlinchey-Yap, *Wintering* by Katherine May, *Radical Acceptance* by Tara Brach, *Wild Mind* by Natalie Goldberg, and *Trust Life* by Louise Hay.

A number of different environments provided nurturing support for me to write parts of this book. Among them are the Mabel Dodge Luhan House, the Ojai Retreat, Madeline Island School of the Arts (MISA), U See I Write retreats at UCI, the online Year-End Writing Retreat with Tara Mohr in 2020 (and its writing prompt "I never thought I was located here"), the Sit/Walk/Write/Read/Listen online intensive practice period with Natalie Goldberg and Rob Wilder, and monthly sitting and sharing online with Natalie Goldberg from 2021 to 2022.

Thank you to the participants of the MISA Writers' Salon with Natalie from July 5 to 9, 2021, for helpful feedback and support while sharing early portions of the book.

Garland-Rose, thank you for making me a mom, and for showing me every day what a life lived fully in the moment looks like. My mom was right. Our children are our teachers. You continue to be mine. I love you to the Moon and stars and galaxies and back.

Steven, you and Garland-Rose are my everything. It all started with you. Thank you for all you did that allowed me the time to write this book, and all that we experienced together that populated much of its contents. This book is for us.

MENTAL HEALTH RESOURCES

The following resources offer hope to those struggling with relationships, money, anxiety, stress, isolation, racism. The descriptions are from the sites themselves.

AL-ANON
al-anon.org

Al-Anon is a mutual support program for people whose lives have been affected by someone else's drinking. By sharing common experiences and applying the Al-Anon principles, families and friends of alcoholics can bring positive changes to their individual situations, whether or not the alcoholic admits the existence of a drinking problem or seeks help.

Alateen, a part of the Al-Anon Family Groups, is a fellowship of young people (mostly teenagers) whose lives have been affected by someone else's drinking whether they are in your life drinking or not. By attending Alateen, teenagers meet other teenagers with similar situations. There are no dues or fees in Al-Anon and Alateen meetings.

DEBTORS ANONYMOUS
debtorsanonymous.org

Debtors Anonymous offers hope for people whose use of unsecured debt causes problems and suffering in their lives and the lives of others. It is supported solely

MENTAL HEALTH RESOURCES

through contributions made by members; there are no dues or fees. Members find relief by working the D.A. recovery program based on the Twelve-Step principles.

Business Debtors Anonymous (B.D.A., formerly known as Business Owners Debtors Anonymous, or B.O.D.A.) is a distinct and dynamic but not separate part of D.A., created to focus on the recovery of members of the fellowship who are business owners. Together, members of B.D.A. support one another in applying the D.A. principles and tools when owning and running a business.

Insight Timer
insighttimer.com

The #1 free app for sleep, anxiety, and stress.

Calm Blog
calm.com/blog

Dedicated to adventures in mindfulness, the power of a good night's sleep, and cultivating a healthier and happier life.

Kaiser Permanente Mental Health and Wellness Tools
healthy.kaiserpermanente.org/southern-california/health-wellness/mental-health/tools-resources

Take a moment. Take a breath. Take time for self-care. Explore our broad range of self-care resources—including apps, audio activities, articles, and more—designed to help you thrive in mind, body, and spirit.

Mental Health America
mhanational.org

Mental Health America is the nation's leading community-based nonprofit dedicated to addressing the needs of those living with mental illness and promoting the overall mental health of all.

National Center for Faculty Development & Diversity (NCFDD)

facultydiversity.org

NCFDD provides on-demand access to the mentoring, tools, and support needed to be successful in the Academy. We focus on four key areas that help achieve extraordinary writing and research productivity while maintaining a full and healthy life off campus: strategic planning, explosive productivity, healthy relationships, and work-life balance.

Harvard Faculty of Arts and Sciences Anti-Racism Resources

projects.iq.harvard.edu/antiracismresources

In the current climate of racial tension and police brutality, it is quite easy to feel overwhelmed by the onslaught of heartbreaking news and information. Yet through the whirlwind of chaos, change in the system is occurring and now more than ever, people are vocal on prevalent issues of racism, encouraging others to join in the fight against systemic racism. However, simply not being a racist is insufficient in eradicating the problem. We must work on actively becoming Anti-Racist in order to properly push back against the system that oppresses Black, Indigenous, People of Color (BIPOC). Members of our community have sought out and compiled resources that can educate, facilitate, and equip those seeking to become more effective anti-racism allies. We hope that these resources will prove helpful in the journey toward a more equal, united America.

Black in Astro

blackinastro.com

Celebrating and amplifying the Black experience in space-related fields. The Black in Astro Community formed through a need for support and community among predominantly early-career Black people working in astronomy. Black in Astro has grown to have members across the globe.

Rising Stargirls

risingstargirls.org

Rising Stargirls is dedicated to encouraging girls of all colors and backgrounds to learn, explore, and discover the universe. We do this by engaging girls in interactive astronomy workshops using theater, writing, and visual art to address each girl as a whole. This provides an avenue for individual self-expression and personal exploration that is interwoven with scientific engagement and discovery. We are committed to the idea that there is no one way to be a scientist, and that together, both science and the arts can create enlightened future scientists and imaginative thinkers.

Vanguard STEM

vanguardstem.com

Vanguard: Conversations with Women of Color in STEM, or #VanguardSTEM for short, is an online platform and community that centers the experiences of women of color, girls of color and non-binary people of color in STEM. Our programming centerpiece is our live webseries. It's a lively gathering moderated by founder and host, Dr. Jedidah Isler, with questions and input from our viewers via social media. *The guiding principle of the show is to create conversations between emerging and established women of color in STEM, where we can celebrate and affirm our identities and STEM interests in a safe space.*

League of Underrepresented Minoritized Astronomers (LUMA)

sites.bu.edu/luma

The League of Underrepresented Minoritized Astronomers (LUMA) is a peer mentoring community for those who self-identify as Black/Indigenous/Latinx women and who are graduate students, postdoctoral researchers, faculty, and research scientists in physics, astronomy, and planetary science. LUMA was founded in 2015 by Prof. Catherine Espaillat, who also serves as LUMA Director. Many Black/Indigenous/Latinx women experience isolation due to our minoritized status in academic spaces. LUMA's goal is to provide a safe, supportive virtual community where we belong and can shine.

FURTHER READING

The analogy I included at the beginning of the book for the number of stars in the observable universe came from the introductory textbook *Life in the Universe* by Jeffrey Bennett and Seth Shostak. Books like these are great for readers, young and old, who are new to the field of astrobiology.

A book I discovered as a child and carried around with me so I could teach myself astronomy while I waited for the opportunity to take classes in school is *Astronomy: A Self-Teaching Guide* by Dinah L. Moché. It's now in its eighth edition.

The book *Planetary Climates* by Andrew Ingersoll offers a great introduction to the topic without too many equations to overwhelm a beginner. You will learn why the climates of the planets in our solar system are the way they are, and step out onto the end of a springboard to look at what the climates of planets around other stars might be like.

There are also wonderful websites for learning and asking questions about the universe, and exoplanets in particular, including Ask an Astronomer (curious.astro.cornell.edu), Exoplanet Exploration (exoplanets.nasa.gov), and Astrobiology at NASA (astrobiology.nasa.gov). Astronomy Picture of the Day (apod.nasa.gov/apod/astropix.html) shows you a different image or photograph of the universe every single day of the year.

There is a wonderful podcast called The Limit Does Not Exist devoted to human Venn diagrams. Its cofounder Christina Wallace authored *The Portfolio*

FURTHER READING

Life (Hachette), a playbook to build a diversified, flexible life that incorporates all of the pieces of your identities.

Remember, there is no one way to be a scientist. Or an artist, for that matter. If you're curious and creative, you qualify.

Keep looking up!

BIBLIOGRAPHY

Abbot, Dorian S., and Raymond T. Pierrehumbert. "Mudball: Surface Dust and Snowball Earth Deglaciation." *Journal of Geophysical Research: Atmospheres* 115, no. D3 (2010). https://doi.org/10.1029/2009JD012007.

African American Women in Physics. "The Physicists." aawip.com/aawip-members.

Baca, Jimmy S. "A Daily Joy to Be Alive." PoemHunter.com. poemhunter.com/poem/a-daily-joy-to-be-alive.

———. *A Place to Stand*. New York: Grove Press, 2002.

Beach, Sarah R. *Mindful Moments for Busy Moms*. New York: Ryland Peters & Small, 2018.

Bennett, Jeffrey, and Seth Shostak. *Life in the Universe*. 3rd ed. New York: Pearson, 2011.

Bochanski, John J., Suzanne L. Hawley, et al. "The Luminosity and Mass Functions of Low-Mass Stars in the Galactic Disk. II. The Field." *The Astronomical Journal* 139, no. 6 (June 2010). https://doi.org/10.1088/0004-6256/139/6/2679.

Brach, Tara. *Radical Acceptance*. New York: Random House, 2004.

Brown, Margaret W. *Goodnight Moon*. New York: HarperCollins, 2007.

Desch, S. J., and A. P. Jackson. "1I/'Oumuamua as an N_2 Ice Fragment of an Exo-Pluto Surface II: Generation of N_2 Ice Fragments and the Origin of 'Oumuamua." *Journal of Geophysical Research: Planets* 126, no. 5 (2021): e2020JE006807. https://doi.org/10.1029/2020JE006807.

Dove, Rita. *Selected Poems of Rita Dove*. New York: Vintage, 1993.

Fraknoi, Andrew. "How Fast Are You Moving When You Are Sitting Still?" *The Universe in the Classroom* 71 (Spring 2007).

Gilbert, Elizabeth. *Eat Pray Love*. New York: Riverhead Books, 2007.

Gillon, Michaël, Amaury H. M. J. Triaud, et al. "Seven Temperate Terrestrial Planets around the Nearby Ultracool Dwarf Star TRAPPIST-1." *Nature* 542 (2017). https://doi.org/10.1038/nature21360.

Goldberg, Natalie. *Wild Mind*. New York: Bantam, 1990.

BIBLIOGRAPHY

Gurian, Anita. "How to Raise Girls with Healthy Self-Esteem." Fox Meadow School. https://www.scarsdaleschools.k12.ny.us/site/default.aspx?PageType=3&ModuleInstanceID=19176&ViewID=7b97f7ed-8e5e-4120-848f-a8b4987d588f&RenderLoc=0&FlexDataID=21913&PageID=17262.

Hagen, Uta. *A Challenge for the Actor.* New York: Charles Scribner's Sons, 1991.

Hay, Louise. *Trust Life.* Carlsbad, CA: Hay House, 2018.

Henley, William Ernest. "Invictus." *Words for the Year* (blog), April 28, 2014. wordsfortheyear.com/2014/04/28/invictus-william-ernest-henley.

Hunter, Deidre A., Bruce G. Elmegreen, and Aomawa L. Baker. "The Relationship between Gas, Stars, and Star Formation in Irregular Galaxies: A Test of Simple Models." *The Astrophysical Journal* 493, no. 2 (1998). https://doi.org/10.1086/305158.

Ingersoll, A. *Planetary Climates.* Princeton, NJ: Princeton University Press, 2013.

Joshi, Manoj M., and Robert M. Haberle. "Suppression of the Water Ice and Snow Albedo Feedback on Planets Orbiting Red Dwarf Stars and the Subsequent Widening of the Habitable Zone." *Astrobiology* 12, no. 1 (2011). https://doi.org/10.1089/ast.2011.0668.

Leiber, Fritz. "A Pail of Air." In *Selected Stories.* San Francisco, CA: Night Shade, 2010.

Levertov, Denise. *Breathing the Water.* New York: New Directions, 1987.

Lobo, Ana, Aomawa L. Shields, Igor Palubski, and Eric T. Wolf. "Terminator Habitability: the Case for Limited Water Availability on M-Dwarf Planets." *The Astrophysical Journal*, in review.

Lorde, Audre. "Smelling the Wind." In *The Collected Poems of Audre Lorde.* New York: W. W. Norton & Company, 2000.

Magee, John G., Jr. "High Flight." "Letter to Parents," September 3, 1941. John Magee Papers, Library of Congress, Washington, DC. Manuscript. In *Respectfully Quoted: A Dictionary of Quotations Requested from the Congressional Research Service*, ed. Suzy Platt (Washington, DC: Library of Congress, 1989), 117–18.

May, Katherine. *Wintering.* New York: Riverhead Books, 2020.

McGlinchey-Yap, Kristine. *Momtras.* Independently published, 2020.

Moché, Dinah L. *Astronomy: A Self-Teaching Guide*, 8th ed. Hoboken, NJ: Wiley, 2014.

Mohr, Tara. *Playing Big.* New York: Avery, 2015.

Morand, Michael. "1917 NAACP Silent Protest Parade, Fifth Avenue, New York City," July 26, 2020. https://beinecke.library.yale.edu/1917NAACPSilentProtestParade.

National Park Service. "1913 Woman Suffrage Procession." https://www.nps.gov/articles/woman-suffrage-procession1913.htm.

Oliver, Mary. "The Journey." In *Dream Work.* New York: Atlantic Monthly Press, 1986.

———. "The Summer Day." In *Dream Work.* New York: Atlantic Monthly Press, 1986.

Palubski, Igor Z., Aomawa L. Shields, and Russell Deitrick. "Habitability and Water Loss Limits on Eccentric Planets Orbiting Main-Sequence Stars." *The Astrophysical Journal* 890, no. 30 (2020). https://doi.org/10.3847/1538-4357/ab66b2.

Pierrehumbert, R. T. "A Palette of Climates for Gliese 581g." *The Astrophysical Journal Letters* 726, no. 1 (2011). https://doi.org/10.1088/2041-8205/726/1/L8.

BIBLIOGRAPHY

Pierrehumbert, R. T., D. S. Abbot, A. Voigt, and D. Koll. "Climate of the Neoproterozoic." *Annual Review of Earth and Planetary Sciences* 39 (2011). https://doi.org/10.1146/annurev-earth-040809-152447.

Procter, Adelaide A. "A Legend of Provence" (1859). Poetry Nook. poetrynook.com/poem/legend-provence.

Rakow, Susan R. "Middle Matters: Guiding Gifted Girls through the Middle School Maze." *Newsletter of the National Association for Gifted Children* (1998).

Rockquemore, Kerry A., and Tracey Laszloffy. *The Black Academic's Guide to Winning Tenure—Without Losing Your Soul*. Boulder, CO: Lynne Rienner Publishers, 2008.

Rogers, Leslie A. "Most 1.6 Earth-Radius Planets Are Not Rocky." *The Astrophysical Journal* 801, no. 41 (2015). https://doi.org/10.1088/0004-637X/801/1/41.

Rushby, Andrew J., Aomawa L. Shields, and Manoj Joshi. "The Effect of Land Fraction and Host Star Spectral Energy Distribution on the Planetary Albedo of Terrestrial Worlds." *The Astrophysical Journal* 887, no. 1 (2019). https://doi.org/10.3847/1538-4357/ab4da6.

Rushby, Andrew J., Aomawa L. Shields, Eric T. Wolf, Marysa Laguë, and Adam Burgasser. "The Effect of Land Albedo on the Climate of Land-Dominated Planets in the TRAPPIST-1 System." *The Astrophysical Journal* 904, no. 2 (2020). https://doi.org/10.3847/1538-4357/abbe04.

Sagan, Carl. *Contact*. New York: Pocket Books, 1997.

Sagan, Carl, and Ann Druyan. *The Demon-Haunted World: Science as a Candle in the Dark*. New York: Random House, 1997.

Shields, Aomawa. "Claiming a Louder Life: Part I." *Inside Higher Ed*, February 19, 2021. insidehighered.com/advice/2021/02/19/remaining-publicly-silent-face-overtly-racist-and-exclusionary-attitudes-has-cost.

———. "Claiming a Louder Life: Part II." *Inside Higher Ed*, March 5, 2021. insidehighered.com/advice/2021/03/05/professor-describes-how-she-found-her-voice-speak-truth-about-systemic-racism-and.

———. "How We'll Find Life on Other Planets." Filmed January 2016. TED Talks video. ted.com/talks/aomawa_shields_how_we_ll_find_life_on_other_planets?language=en.

———. "Nefertiti with a Calculator." *Bricolage Literary and Arts Journal* (2011): 59.

———. "Should We Be Looking for Life Elsewhere in the Universe?" TEDEd. ed.ted.com/lessons/should-we-be-looking-for-life-elsewhere-in-the-universe-aomawa-shields.

Shields, Aomawa L. "The Climates of Other Worlds: A Review of the Emerging Field of Exoplanet Climatology." *The Astrophysical Journal* Supplement Series 243, no. 2 (2019). https://doi.org/10.3847/1538-4365/ab2fe7.

———. *The Effect of Star-Planet Interactions on Planetary Climate*. Doctoral thesis, University of Washington, 2014. ui.adsabs.harvard.edu/abs/2014PhDT.......634S/abstract.

———. "Space Was Her Second Act: As Spitzer Takes Its Final Bow, Astronomer and Actress Aomawa Shields Thanks Spitzer For Her Big Break." Jet Propulsion Laboratory, Spitzer Space Telescope, October 23, 2019. https://www.spitzer.caltech.edu/blog/space-was-her-second-act-as-spitzer-takes-its-final-bow-astronomer-and-actress-aomawa-shields-thanks-spitzer-for-her-big-break.

BIBLIOGRAPHY

Shields, Aomawa L., Sarah Ballard, and John Asher Johnson. "The Habitability of Planets Orbiting M-dwarf Stars." *Physics Reports* 663 (2016): 1–38. https://doi.org/10.1016/j.physrep.2016.10.003.

Shields, Aomawa L., Rory Barnes, Eric Agol, Benjamin Charnay, Cecilia Bitz, and Victoria S. Meadows. "The Effect of Orbital Configuration on the Possible Climates and Habitability of Kepler-62f." *Astrobiology* 16, no. 6 (2016). https://doi.org/10.1089/ast.2015.1353.

Shields, Aomawa L., Cecilia Bitz, and Igor Palubski. "Energy Budgets for Terrestrial Extrasolar Planets." *The Astrophysical Journal Letters* 884, no. 1 (2019). https://doi.org/10.3847/2041-8213/ab44ce.

Shields, Aomawa L., Cecilia M. Bitz, Victoria S. Meadows, Manoj M. Joshi, Tyler D. Robinson. "Spectrum-Driven Planetary Deglaciation Due to Increases in Stellar Luminosity." *The Astrophysical Journal Letters* 785, no. 1 (2014). https://doi.org/10.1088/2041-8205/785/1/L9.

Shields, Aomawa L., and Regina C. Carns. "Hydrohalite Salt-albedo Feedback Could Cool M-Dwarf Planets." *The Astrophysical Journal* 867, no. 1 (2018). https://doi.org/10.3847/1538-4357/aadcaa.

Shields, Aomawa L., Victoria S. Meadows, Cecilia M. Bitz, Raymond T. Pierrehumbert, Manoj M. Joshi, and Tyler D. Robinson. "The Effect of Host Star Spectral Energy Distribution and Ice-Albedo Feedback on the Climate of Extrasolar Planets." *Astrobiology* 13, no. 8 (2013). https://doi.org/10.1089/ast.2012.0961.

Shikibu, Izumi. "Watching the Moon." In *The Ink Dark Moon: Love Poems by Ono no Komachi and Izumi Shikibu, Women of the Ancient Court of Japan*, translated by Jane Hirshfield and Mariko Aratani. New York: Vintage, 1990.

Warren, Stephen G. "Optical Properties of Ice and Snow." *Philosophical Transactions of The Royal Society A* 377, no. 2146 (2019). https://doi.org/10.1098/rsta.2018.0161.

Weir, Kirsten. "Old Problem, Old Solutions." *Spectrum* 35 (2007). aas.org/sites/default/files/2019-09/spectrum_Jun07.pdf.

Williams, Chad. "100 Years Ago African-Americans Marched Down 5th Avenue to Declare that Black Lives Matter." The Conversation, July 25, 2017. https://theconversation.com/100-years-ago-african-americans-marched-down-5th-avenue-to-declare-that-black-lives-matter-81427.

Wolf, Eric T., Aomawa L. Shields, Ravi K. Kopparapu, Jacob Haqq-Misra, and Owen B. Toon. "Constraints on Climate and Habitability for Earth-like Exoplanets Determined from a General Circulation Model." *The Astrophysical Journal* 837, no. 2 (2017). https://doi.org/10.3847/1538-4357/aa5ffc.

Woods, Paul, and Ashley Walker. "The Representation of Blackness in Astronomy." *Nature Astronomy* 6 (2022). https://doi.org/10.1038/s41550-022-01704-0.

Zhou, Chuanming, Magdalena H. Huyskens, Xianguo Lang, Shuhai Xiao, and Qing-Zhu Yin. "Calibrating the Terminations of Cryogenian Global Glaciations." *Geology* 47, no. 3 (2019). https://doi.org/10.1130/G45719.1.